The True North

Also by D. Paul Schafer from Rock's Mills Press

The Age of Culture (2014)
The Secrets of Culture (2015)
Celebrating Canadian Creativity (2016)
Will This Be Canada's Century? (2018)
The Cultural Personality (2018)

The True North
How Canadian Creativity Changed the World

D. Paul Schafer

Rock's Mills Press
Oakville, Ontario
2019

Published by
Rock's Mills Press
www.rocksmillspress.com

Copyright © 2019 by D. Paul Schafer.
All rights reserved. Published by arrangement with the author.

For information, contact Rock's Mills Press at
customer.service@rocksmillspress.com.

Contents

Preface ... *vii*

1. A Remarkable Discovery ... *1*

2. Transportation and Communications ... *18*

3. Resources, Agriculture, and Industry ... *65*

4. Medicine, Health Care, and Compassion ... *90*

5. Arts and Entertainment ... *121*

6. Activism and Advocacy ... *152*

7. Sports and Recreation ... *198*

8. Education, Politics, and Political Milestones ... *235*

9. Looking Back, Going Forward ... *271*

Selected Readings ... *281*

Index ... *283*

A section of illustrations will be found between pages 120 and 121, following Chapter Four.

Preface

This is a book about how Canadian creativity has changed the world. I have written it because I have discovered during the course of my research on Canada and Canadian culture that Canadians are very creative people and their creativity has had a major impact on the world.

This is not the first book I have written on Canadian creativity. The first one—*Celebrating Canadian Creativity*—was written several years ago and documented the fact that Canadians have been creative throughout their history in virtually every aspect of the country's cultural life, from food, clothing, shelter, transportation, and communications to the arts, sciences, sports, recreation, and the environment. This creativity has come from women and men, small towns and big cities, rural and urban areas, and recent immigrants as well as long-time residents. The focus in this book was on the people themselves, their creative achievements, when these achievements first occurred, and how they affected Canada and Canadians.

The present book is different, although it shares certain similarities with the first. Whereas the first book was designed to show that Canadians are very creative and document how their creativity has impacted on the country and its citizenry, *The True North* is, as its subtitle notes, designed to show how Canadian creativity has changed the world. In order to best do this, I have limited my coverage of the creative contributions of Canadians in this book to those I think have had the greatest impact on the world. This has made it possible to deal with these contributions in much more depth, especially with respect to how they have spread throughout the world and changed it for the better. For only in this way is it possible to document the powerful effect Canadian creativity has had on the world, as well as to confirm how this has changed the world in some very fundamental and positive ways—and how it may change the world to an even greater extent in the years ahead.

This book would not have been possible without contributions from family members and many close friends and colleagues. First of all, I would like to thank my wife Nancy and daughters Charlene and Susan, as well as my brother Murray and sister-in-law Eleanor James, for their support and encouragement with this book. I would also like to thank John Hobday, Walter Pitman, James Gillies, André Fortier, Joyce Zemans, Don McGregor, John Hawco, Ian Morrison, Barry Witkin, Arthur Witkin, John Gordon, Frank Pasquill, Peter Sever, and others for their assistance and companionship during my efforts to expand knowledge and understanding of the crucial role Canadian creativity has played in global development and world affairs. I would especially like to express my gratitude to Bill Thachuk for the valuable contributions he made to the research and writing of this book. Finally, and most of all, I would like to thank my close friend and publisher, David Stover, for his interest in my writings on the subject of culture in general and Canadian culture and creativity in particular, as well as for his commitment to publishing a number of my books on these matters over the last five years.

D. Paul Schafer
Markham, Canada
2019

The True North

Chapter One
A Remarkable Discovery

It all began one evening in late November more than twenty-five years ago. With a great deal of spare time on my hands and a long Canadian winter to endure, I decided to read the *Canadian Encyclopedia*. Can you believe it? Who in their right mind would do something as challenging—or as foolish—as this?

When I say "read" the *Canadian Encyclopedia*, I really mean that I read and scanned the *Encyclopedia*. Most of the entries and sections that were of great interest to me and that I thought would be helpful in my work I read word by word and line by line. However, I scanned some entries and sections that were of less interest and not relevant to my work. My purpose was to acquire as much knowledge, understanding, information, and ideas as I could on Canada, Canadians, and Canadian culture, and I felt that working my way through the *Canadian Encyclopedia* was the best way to achieve this. So there I was on that first evening in late November many years ago with a pencil in one hand and a note pad on the table prepared to work my way through the *Canadian Encyclopedia* from beginning to end, stopping to make notes whenever I came across something very interesting or that I wanted to record for future use.

It wasn't long before I began to get the impression that Canadians are incredibly creative. I also began to get the impression that their creativity has benefitted not just Canada and Canadians, but also people and countries in other parts of the world and the world as a whole. This came as a real surprise! Apart from a few well-known examples of Canadian creativity—such as the invention of the telephone, the discovery of insulin, the establishment of standard time, and the creation of basketball—I was quite unaware of this. I suspected that this must be true for most other Canadians, as well as for people and countries in other parts of the world.

On the very second page, for instance, I came across an entry for *Abbott, Maude Elizabeth Seymour*. This inventive woman was born in St. Andrews East in Quebec in 1869 and graduated from McGill University in general arts. When she was barred from studying medicine at McGill University because she was a woman, she went to Bishop's University in Lennoxville, Quebec to obtain her medical degree. I liked her instantly for this.

Abbott eventually went on to become the curator of the McGill Medical Museum and achieve international recognition for her creative contributions and outstanding research with respect to "blue babies" and congenital heart disease. She quickly became well-known in medical circles throughout the world, especially when she served as permanent secretary of the International Association of Medical Museums and was editor of its journal from 1907 to 1938.

The next entry that caught my eye was on *Adult Education* a few pages later. While I was unaware of it, I discovered that Canadians have been actively involved in adult education since Confederation in 1867, with strong contributions coming from mechanical institutes in Ontario, Quebec, and Nova Scotia, Queen's University through its extramural and extension programs, the YMCA through its night classes for the military, the Canada Farmers' Institutes, and the National Council of Women. This was also true of the Women's Institutes across the country, Frontier College, which was created in 1899 to provide adult learning opportunities for people in remote areas such as logging, mining, and railroad camps, the University of Alberta and the University of Saskatchewan through their extension courses, and Saint Francis Xavier University through its cooperative programs and the world-famous Antigonish Movement created by James Tompkins and Moses Coady.

After reading this, I came across the following statement in the same entry: "Internationally, Canadians contributed to the development of adult education through such organizations as the International Congress of University Continuing Education, and UNESCO. With support of participants from the UNESCO conference in Tokyo (1972), and under the leadership of a Canadian adult educator, Dr. J. Roby Kidd, the International Council for Adult Education (ICAE) was created in 1973."

This Council was situated in Canada as a result of the far-sightedness and international reputation of Dr. Kidd, who is deemed by many educators to be "the father of adult education" in the world. This accolade did not come lightly or out of the blue. Roby Kidd was building on Canada's solid international reputation in adult education that stretched back to Confederation, although he carried this tradition much farther through his pioneering efforts and innovative work in this area in Canada and other parts of the world.

As I made my way through the pages following the entry on Adult Education, I was drawn to a picture and write-up on *Aitken, William Maxwell, 1st. Baron Beaverbrook*. Son of a Presbyterian minister, Aitken was born in Maple, Ontario in 1879 and amassed a huge fortune in Canada by selling stocks and bonds. He moved to England in 1910 where he became actively involved in British politics and the newspaper industry. He was knighted by King George V in England in 1911 and was known thereafter as Lord Beaverbrook. He went on to represent Canada and the Canadian government on the military front during World War I and wrote *Canada in Flanders*. After the war, Aitken acquired a chain of newspapers in Britain including the *Daily Express* and *Evening Standard* as well as founding the *Sunday Express*. This made him one of the most powerful people in Great Britain and the newspaper business. He also served as a minister of aircraft production in Churchill's government during World War II and became a major benefactor of the arts and education later in life, especially with respect to the Beaverbrook Art Gallery and University of New Brunswick in Canada.

As my reading progressed on that first evening, I was attracted to an entry on *Anka, Paul Albert*. While present generations of Canadians and people in other parts of the world are probably not aware of it, Paul Anka was one of the greatest song writers and popular singers in the world in the middle part of the twentieth century, much like Drake, Michael Bublé, Céline Dionne, Shania Twain, Justin Bieber, and many other Canadian singers are today. Of Lebanese-Canadian descent, Anka became world famous when he was still in his teens for writing and singing such popular songs as "Diana"—the second-highest selling record in history when the *Canadian Encyclopedia* was published—"You Are My Destiny," "Puppy Love," and "Lonely Boy." And this is not all. He also wrote "My Way" and "She's a Lady," made famous throughout the world by Frank Sinatra and Tom Jones respectively. He also wrote the theme for *The Tonight Show with Johnny Carson* on television.

On the very next page, I came across an entry for *Anne of Green Gables*. This classic novel about a little orphan girl named Anne Shirley is well known in Canada and many other parts of the world. It was written by Lucy Maud Montgomery and was first published in Boston in 1908. This proved to be the springboard for several other novels about Anne Shirley, who was adopted in the first novel by Matthew and Marilla Cuthbert and grew up on a farm in Avonlea near Cavendish, Prince Edward Island. The original novel has gone through many editions and been translated into numerous languages, as well as being made into a movie, a musical, and many television programs. It is a favourite story of many children in the world, especially in Japan where it has achieved a cult-like status and huge following. I wonder how many people from Japan have come to Prince Edward Island since *Anne of Green Gables* was published looking for the place where their beloved Anne Shirley grew up and had most of her exciting adventures.

The entry on *Anne of Green Gables* ended my first evening of reading the *Canadian Encyclopedia*. It proved to be very informative. I had only reached page 61 and already I had come across a number of Canadians who had a significant impact on other parts of the world and not just Canada through their creativity. I had the sense that I was on to something very powerful here, and was looking forward to sitting down at the next opportunity to pick up where I left off.

That happened a few evenings later and proved to be a repeat of the first evening. Shortly after reading about *Anne of Green Gables*, I came across an entry for *Atwood, Margaret Eleanor*. Like Lucy Maud Montgomery a hundred years earlier, Margaret Atwood was at that time—and far more so today—well-known throughout the world as one of Canada's and the world's most celebrated and prolific authors. This is due to the many coveted national and international awards she has won and the fact that a number of her books have been made into movies and television series. She has received numerous international and national awards, such as the Man Booker Prize, Arthur C. Clarke Award, Governor General's Award, Franz Kafka Prize, and the National Book Critics and PEN Center USA Lifetime Achievement awards. Many of her books and novels examine feminist themes, mythological matters, and social and human issues, such as *The Edible Woman, The Blind Assassin, The Handmaid's Tale, Cat's Eye,* and *Alias Grace.* The television series based on her novel *The Handmaid's Tale* has been widely acclaimed internationally and won eight Primetime Emmy awards in 2017-2018, including Outstanding Drama Series, as well as a Golden Globe award for best dramatic TV series in 2018. *Alias Grace* also came out in 2018 as a Netflix miniseries, and the *Handmaid's Tale* was extended by Hulu for a third season which premiered in 2019.

The entry on Margaret Atwood was followed by one on the *Avro Arrow (CF-105)* a couple of pages later, as well as one on the *Avro Jetliner (C-102)*. The Avro Arrow was the fastest plane in the world for a time and was described as "an advanced, supersonic, twin-engine, all-weather interceptor jet fighter developed by A. V. Roe of Canada from 1949 until the federal government's controversial cancellation of the project in 1959." It was built with the intention of countering the threat of Soviet bombers over Canada's northern air space during the Cold War. The Avro Jetliner, like the Avro Arrow, was also extremely fast in its day. It was designed by James Floyd and others and flown for the first time on August 10, 1949 when it exceeded a speed of 800 kilometres an hour. It was the first flight of a jet transport in North America and the second in the world by less than two weeks.

Who would have guessed that Canadians would have been

involved in the design and construction of some of the fastest airplanes in the world? Certainly not me, although I was surprised to learn later as I delved more deeply into this subject that Canada has a long and distinguished tradition in aviation and aerospace activity. I also learned as I worked my way through the *Canadian Encyclopedia* that things like this seldom occur out of the blue, as there were earlier creative contributions to aviation by Alexander Graham Bell, Mabel Bell, F. W. "Casey" Baldwin, and J. A. Douglas McCurdy. They were involved in building and flying the Silver Dart in Baddeck, Nova Scotia, the first airplane to be flown in Canada and the Commonwealth, as well as many other Canadian achievements in this area that have had an important impact on the world in the many years since that time.

When I was finished with the entries beginning with the letter "A," I began reading those beginning with "B." This proved to be a repeat of the first section, since it was filled with a great deal of information about many other creative achievements by Canadians that have had a significant impact on the world and not just on Canada.

The first entry beginning with "B" that caught my attention was one for *Back, Frédéric*. He was born in France in 1924 and attended the École des Beaux-Arts in Rennes before coming to Canada in 1948. A talented film maker, graphic designer, artist, and educator, Back was hired by the CBC in Montreal and created many popular animation films for children when he was there, eventually winning an Academy Award Oscar for *Crac* in 1982. While I was only vaguely aware of it at the time, I discovered later that Canada already had an outstanding international reputation in film animation when Back won this award. This was due largely to Norman McLaren, one of the world's greatest and most creative film animators when he was at the National Film Board of Canada in the 1940s and '50s. His pioneering work in film animation made a valuable contribution to many animators who followed in his footsteps in Canada and other parts of the world, as well as having an influence on Walt Disney and some of his animation films.

Several pages later, I came across an entry for *Ballet*. After reading about the historical development of ballet, I was surprised to learn that several Canadian ballet dancers—including Patricia Wilde from Ottawa, Melissa Hayden from Toronto, and Lynn Seymour from Alberta and later Vancouver—achieved international recognition with well-known American and British ballet companies in the middle part of the twentieth century because Canada had no major ballet companies of its own at that time. This was set right a decade or two later when the Royal Winnipeg Ballet, the National Ballet of Canada, and Les Grands Ballet Canadiens were created.

These three companies were created by women who went on to achieve international prominence and status for their accomplishments. Gweneth Lloyd and Betty Farrally played a seminal role in forming the Royal Winnipeg Ballet; Celia Franca played a similar role by founding the National Ballet of Canada in Toronto; and Ludmilla Chiriaeff created Les Grands Ballet Canadiens in Montreal. All three women came to Canada from elsewhere in the world and along with many Canadians built these companies into international powerhouses. They did so by training internationally-known Canadian ballet dancers such as Karen Kain, Veronica Tenant, Evelyn Hart, Frank Augustine, and many others, as well as enticing some of the world's greatest ballet dancers to come to Canada and dance with Canadian companies, most notably Rudolf Nureyev and

Mikhail Baryshnikov from the USSR and Erik Bruhn from Denmark.

Shortly after reading about Canada's creative achievements in ballet and the international consequences of this, I came across an entry for *Banting, Sir Frederick Grant.* Banting was one of the co-discoverers of insulin along with Charles Best, J. J. R. Macleod, and J. S. Collip. A phenomenal achievement by any standard, the discovery of insulin in the 1920s provided a treatment for diabetes, one of the most deadly and dreadful diseases in the world. Not only was this a highly creative discovery, but also it has saved millions of lives in Canada and elsewhere in the world since that time, something that has become extremely important as the incidence of diabetes is on the rise in many parts of the world due to increases in the consumption of sugar and sugar-based products, as well as the lack of physical exercise, poor diets, obesity, and other factors.

The next entry that caught my attention a few pages later was *Basketball.* This is a sport where Canadians have made one of their most remarkable creative contributions in an international sense. The man who created basketball, James A. Naismith, was a Canadian who grew up in Almonte near Ottawa and attended McGill University. Following this, he got a job at the YMCA International Training School in Springfield, Massachusetts in the United States. He invented basketball when he was there because the director at the Training School desperately needed a game that young people and students could play in the winter—they had baseball, football, and other sports in the spring, summer, and fall—and Naismith was charged with this responsibility. The result was basketball, with thirteen rules handwritten by Naismith and the use of peach baskets for hoops. Since that time, basketball has spread like wildfire throughout the world and gone on to become one of the most widely played and watched sports in the world.

What piqued my interest in this achievement even more was a reference that was made in the same entry to Canada having one of the greatest basketball teams in the world at one time. It was the Edmonton Grads, a women's basketball team that won four world championships and had a record of 502 wins in 522 games—including 78 consecutive wins—between 1915 and 1940. Despite this achievement, the game has been dominated primarily by Americans since that time.

The entry on basketball was followed by one on *Bata, Thomas John* on the very next page. Thomas John Bata was the son of Thomas Bata who built and owned what was the largest shoe manufacturing company in the world at one time in Zlin, Czechoslovakia (now the Czech Republic) in the first few decades of the twentieth century. Tom Bata Jr., as he was called, came to Canada in 1939 and built a major plant in the company-owned community of Batawa, Ontario. Although the company was well established in 30 countries throughout the world prior to this time, Tom Bata Jr. expanded the company considerably, so much so that by 1983 the company had factories, stores, and related outlets in 95 countries. It employed some 88,000 people at this time, made and sold more than a million pairs of shoes a day, and was the largest shoe manufacturer in the world, thereby explaining why Tom Bata Jr. became known as "the shoemaker to the world." Creative achievements like this were enhanced even more when Tom's wife, Sonja Bata, created the Bata Shoe Museum in Toronto. It is the largest and most innovative shoe museum in the world with thousands of shoes from all over the world in its vast and varied collection.

Several pages after the Bata entry, I came across one

for *Bell, Alexander Graham*. I knew the name instantly, as most Canadians do, since Bell is a real Canadian hero and the international genius who invented the telephone. What I didn't know, and was surprised to learn, was that Bell was an incredibly creative person with countless other inventions and "firsts" to his credit in addition to his contributions to aviation mentioned earlier. While Bell spent a great deal of time in the United States and eventually became an American citizen, many of his most creative achievements were realized in Canada at his summer residence and research facilities at Baddeck, Nova Scotia. This was where Bell did a great deal of pioneering work on aircraft, the photoelectric cell, the iron lung, the desalination of seawater, the phonograph, and the hydrofoil boat.

The next entry that caught my attention was *Berton, Pierre*. I knew his name instantly as well. He was a well-known journalist, historian, and media personality who was born in Whitehorse, Yukon Territory in 1920 and grew up in Dawson, one of the major centres of the Klondike Gold Rush of 1898. His first important book—*Klondike* (1958)—described the gold rush days. This topic was very appropriate because Berton's father had panned for gold during those days. In the years to follow, Berton manifested a penchant for selecting and writing about Canadian events with major national and international implications and appeal, most notably the building of the Canadian Pacific Railroad in such books as *The National Dream* (1970) and *The Last Spike* (1971), and, exactly ten years later, two major books on the War of 1812 with the United States, *The Invasion of Canada* (1980) and *Flames Across the Border* (1981). While Berton wrote many other books, these books established his reputation as Canada's greatest popular historian, with a real penchant for writing books that were patriotic, filled with colourful images and intimate details, and interesting narratives and ideas. It was Berton who said, "a true Canadian is one who can make love in a canoe without tipping."

The entry on Berton was followed by one on *Bethune, Henry Norman*. He was a medical doctor, surgeon, inventor, and political activist who was born in Gravenhurst, Ontario in 1890 and died at Huang Shiko in North China in 1939. Not only was he the inventor of a number of medical devices, but also he assisted China and the Chinese people in their wars with Japan. Bethune became internationally known when he was hailed by Mao Zedong, China's military leader, who urged all Chinese people—and especially all Chinese communists—to be like Norman Bethune by emulating his humanitarianism, internationalism, commitment to causes, and concern for others.

An entry shortly after the write-up on Bethune also attracted my attention. It was *Bluenose, The*. This is without doubt the most famous Canadian sailing ship of all. I recognized the name at once because I lived in Nova Scotia for several years when I was teaching at Dalhousie University in Halifax and Acadia University in Wolfville. The *Bluenose* was built at the Smith and Rhuland shipyard in Lunenburg, Nova Scotia, a well-known UNESCO world heritage site, which I visited many times. The *Bluenose* became famous throughout the world by beating the fastest schooners from United States and other countries on its way to winning the International Fisherman's Trophy—emblematic of the fastest sailing ship in the world—in 1921, 1922, and 1923 and again in 1931 and 1938. Its replica, the *Bluenose II*, was also built at the Smith and Rhuland shipyard in Lunenburg and continues to sail the high seas and win many international accolades today as one of the most recognizable and popular "tall ships" in the world.

Following the *Bluenose* entry, the next entry that had a strong impression on me was in the field of transportation as well. However, this time it was transportation over snow rather than by water. It was the entry for *Bombardier, J. Armand*.

What an incredibly creative individual Armand Bombardier was! Born in Valcourt, Quebec in 1908, he dreamed of creating vehicles that could travel over snow, inspired no doubt by Quebec's long winters and heavy snowfalls and snowdrifts. This inspiration eventually gave rise to the creation of the snowmobile, the Ski-Doo, and many other recreational, military, and industrial vehicles capable of travelling over snow, beginning in the 1930s and continuing to the present day.

Several pages later, my eyes settled on an entry for *Bowling, Indoor*. Like the entry on *Ballet*, this entry commenced with some general information on the historical development of bowling. It revealed that a number of games similar to bowling were played in Egypt as far back as 5,000 BCE, although 10-pin bowling was created in the United States in the nineteenth century. Nevertheless, the 10-pin game proved too long and arduous for many bowlers, primarily because of the weight of the balls and the number of pins. So Thomas F. Ryan, a Canadian, invented five-pin bowling in Toronto, a variation on the 10-pin game, to please his customers and create a game they could play over the lunch hour. Ryan's innovation proved that Canadians are not only creative at inventing new games—the invention of many board games such as Trivial Pursuit, Pictionary, Yahtzee, and Crokinole confirms this—but also at adapting existing games to their own specific requirements and needs.

With this entry on *Bowling, Indoor*, my reading—and a bit of scanning!—of the *Canadian Encyclopedia* on the second evening was complete. As on the first evening, I had come across many creative achievements by Canadians that I had not been aware of before. What struck me most after these two evenings was how much Canadians have created that has had an important impact on the world and not just on Canada and Canadians.

I was only up to the entries beginning with "C" and already I had discovered that Canadians discovered insulin, invented basketball, the snowmobile, the Ski-Doo, the mobile blood bank, and five-pin bowling, and had made their mark on many other areas of life that have had a significant impact on the world. I also learned that a Canadian was deemed to be one of the major founders as well as "the father of adult education" in the world, and another was known as "the shoemaker to the world." I began to wonder if Canadians were really as conventional, laid-back, and unimaginative as many people thought. I was beginning to have my doubts.

These doubts were put to rest as I plunged more deeply into the *Canadian Encyclopedia* in the winter evenings that followed. What impressed me the most as those evenings piled up, one on top of another, was how creative Canadians actually are and have been over the centuries, as well as how many things Canadians have created or invented that have had a powerful impact on the world. What I discovered those first few evenings turned out to be the tip of a much larger iceberg—an iceberg that was almost completely submerged under water and hidden from view. Even a number of important achievements that were attributed to Americans or people in other parts of the world that have had a powerful impact on the world—such as the invention of the light bulb, the electric oven, and the variable pitch propeller—were either invented by Canadians or Canadians played a highly inventive and pioneering role in their initial development.

By the end of the winter, I had worked my way through the entire *Canadian Encyclopedia*, although I must confess I was scanning almost as many entries as I was reading towards the very end. Nevertheless, I had achieved what I set out to achieve at the beginning of the winter. I had worked my way through the entire *Encyclopedia*—more than 2,000 pages in all—and had learned an enormous amount about Canada, Canadians, Canadian culture, and Canadian creativity. What I discovered in those pages amazed me, even if I didn't realize the full magnitude or significance of this at the time.

Three things stood out in my mind as I reflected on my experience with this in the days and weeks to follow. The first was how incredibly creative Canadians are as a people and how much of their creativity has been manifested in virtually every area of the country's cultural life, all parts of the country, and among all the diverse peoples and groups that make up the Canadian population.

The second was how intimately Canadian creativity is tied up with the country's climate, geography, natural environment, huge size, and distinct and various seasons. This makes "place" a very important factor in determining the specific character of Canadian creativity. It is no coincidence in this regard that Canadians have been especially creative in transportation and communications, a wide range of clothing, shelter, and accommodation, winter sports, recreational activities, environmental conservation initiatives, and so forth. While it is impossible to say whether Canadians are more creative than any other people in the world, what it is possible to say is that the high level of creativity manifested throughout Canada and by Canadians has resulted from the need to come to grips with the colossal size of the country, the cold climate, the challenging geography, the harshness and hardships of the winter, and the inhospitable nature of the natural environment for a significant part of the year.

The third thing that stood out was that Canadians were and are much more effective at creating things than benefitting from them financially. This goes to the heart of the distinction that is often made between "creativity" and "innovation." Whereas creativity is concerned with having a vision or idea and translating that vision or idea into reality in a tangible or concrete way—such as when Joseph Armand Bombardier had a vision or idea of a vehicle that could travel over snow and then turned this into reality by creating the snowmobile and the Ski-Doo—innovation is concerned with capitalizing on creativity by producing goods, services, products, and devices that yield significant economic, industrial, commercial, and financial benefits.

With this in mind, fast forward now some twenty years after I sat down on that first evening in late November to work my way through the *Canadian Encyclopedia*. I was sitting in a library in Markham one day when I happened to come across a book by Arthur Herman called *How the Scots Invented the Modern World: The True Story of How Western Europe's Poorest Nation Created Our World and Everything in It.*

I found this book fascinating because it documented in detail the phenomenal contributions made by Scotland and the Scottish people to the world we are living in at the present time. Herman made this case by demonstrating how creative achievements by Scotland and some incredibly creative Scottish people several centuries ago—most notably Adam Smith,

James Watt, David Hume, John Knox, Robbie Burns, Francis Hutcheson, and others—laid the foundations for the modern world. As Herman stated on the back cover of his book, "when we gaze out on a contemporary world shaped by technology, capitalism, and modern democracy, and struggle to find our place as individuals in it, we are in effect viewing the world as the Scots did."[1]

This book got me thinking about all those evenings I spent during that winter many years earlier working my way through the *Canadian Encyclopedia.* Although I cannot claim that Canadians had done something as profound or powerful as inventing the modern world or anything close to it, I can claim with confidence that Canadians have made many highly creative and original contributions to the world, and that millions of people around the world have benefitted from Canadian creativity over the centuries—and are still benefitting from it today.

I decided then and there to write a book some day on this remarkable feat—and here it is. It is designed to document how Canadian creativity has changed the world in a whole series of dynamic and fundamental ways. When all this information is pulled together and presented in one place, it makes for a fascinating story, a story that I feel Canadians and people in other parts of the world should know a great deal more about. I hope you find this story as captivating and compelling as I did when I was researching and writing this book.

---

1. Arthur Herman, *How the Scots Invented the Modern World: The True Story of How Western Europe's Poorest Nation Created Our World and Everything in It* (New York: Crown Publishers, 2001), Back Cover.

Chapter Two
Transportation and Communications

It is said that transportation and communications are the lifeblood of Canada. This is true in the sense that many creative achievements have been required in these areas to create a colossal country despite incredible odds, move people, products, resources, information, and ideas over short and long distances, maintain contact with people in other parts of Canada and the world, and achieve a high standard of living and excellent quality of life for as many Canadians as possible.

This would not have been possible without countless developments in transportation by land, water, rail, air, over snow, rough terrain, and in space, as well as communication by telephone, telegraph, newspapers, radio, television, wireless technology, and social media. These developments have been needed to link Canada from the Atlantic Ocean in the east to the Pacific Ocean in the west and the Arctic Ocean in the north, as well as to maintain contact with France, England, other parts of Europe, the United States, and the rest of the world. While the impact of some of these developments has been confined primarily to Canada and Canadians, others have had—and continue to have—a powerful impact on other people, countries, continents, and the world at large.

Some of the most creative achievements have occurred in the area of transportation by water. This is because Canada is blessed with millions of lakes and rivers, countless bays, inlets, and harbours, and thousands of kilometres of shoreline that are ideal for this purpose.

The Indigenous peoples were involved in the development of many of the oldest forms of transportation. This is particularly true for the kayak, the umiak, and the canoe. Historical records reveal that the kayak was invented by native Inuit, Aleut, Yupik, and Ainu hunters in the sub-Arctic regions of northeast Asia, Canada, and Greenland thousands of years ago. Extremely light in weight and durable in nature, the kayak was—and still is today in some cases—constructed of stitched seal and other animal skins stretched across a wooden frame, and navigated by one or two paddlers using a double-sided paddle.

While Inuit hunters in the far north were perfecting the kayak, Indigenous peoples farther south were perfecting the canoe. Like the kayak, the canoe can also be traced back to ancient times. However, Canadians made many original contributions to its evolution, and actually did create the birch bark canoe. Light, durable, and easy to carry, the canoe made it possible for the Indigenous peoples and first Europeans to come to grips with a whole host of complicated geographical and environmental problems by facilitating transportation on and between the country's countless lakes and rivers due to its portability, durability, and especially the availability of birch bark in most parts of the country (hence the title of Roy MacGregor's popular book, *Canoe Country: The Making of Canada*).

Without the canoe in general and the birch bark canoe in particular, it would not have been possible for Samuel de Champlain to explore the interior of Canada as far west as Georgian Bay in 1615; Alexander Mackenzie to cross the North American continent and reach the Arctic Ocean by 1789 and the Pacific Ocean in 1793—a full decade ahead of Meriwether Lewis and William Clark's historic expedition across United States to the Pacific and back in 1804; and David Thompson—one of Canada's best known explorers and most accomplished mapmakers the world has ever known—to travel more than 130,000 kilometres and map over 3,800,000 square kilometres in North America.

Without Canada's involvement in the development of the canoe and the kayak, it is highly unlikely that these modes of transportation would be as popular and ubiquitous as they are today throughout the world for a variety of transportation, economic, recreational, and related purposes. This is particularly true of the kayak, which is used in virtually all parts of the world for white water, surfing, and racing activities. Not only do world championships take place in kayak racing over 100, 200, 500, and 1,000 metres, but also there are several events in men's and women's kayak racing for one, two, and four paddlers at the Olympic Games, as well as in canoe racing in slalom and sprint events since 1936.

While the kayak and the canoe are still popular in most parts of Canada and the world today, many of Canada's most creative contributions to water transportation occurred in the Maritimes. In fact, the Maritimes were a hot bed of creativity in this area in the nineteenth century, so much so that it made many powerful and seminal contributions to water transportation throughout the world.

An incredible amount of this creativity went into the design, development, and construction of ships. Not only was there a constant search for new and better ways to move people, products, mail, and resources by water—driven by the desire to capitalize on the revolution going on in steam power and its application to water transportation—but also many of the wars fought during this time were fought by ships at sea rather than armies on land. Moreover, Britain, France, and other European powers were anxious to maintain contact with their colonies and other countries in the world, largely to take advantage of the lucrative trade opportunities that existed in the nineteenth century. Since this was only possible in most cases by water transportation, an enormous amount of time, effort, money, and ingenuity went into developing this specific mode of transportation.

Strategically located between Europe, United States, the Caribbean, and South America, the Maritimes were ideally situated to capitalize on the rapidly expanding flow of goods, supplies, passengers, immigrants, resources, and mail between various parts of the world. As a result, there was a constant flow of traffic in and out of the Maritimes in general and Halifax in particular.

Into this world of rapidly accelerating marine activity in the North Atlantic and growing fascination with steam power and its application to water transportation stepped Samuel Cunard in the early years of the nineteenth century. He came from a long line of sea merchants and ship builders who emigrated originally to what is now the United States from Prussia in 1683. His family remained in the United States until the 1780s and then came to Canada because they were United Empire Loyalists and deeply committed to Great Britain and the British Empire. Their decision was compounded by the fact that their ships were confiscated by the colonies to the south during the American War of Independence.

Abraham Cunard, father of Samuel, was heavily involved in shipping. Shortly after arriving in Canada, he purchased a large tract of land in Halifax and created Cunard's Wharf, running down to Halifax harbour. After working for shipping companies in Halifax and Boston for several years, Samuel joined the family business on a full-time basis. By 1814, he had acquired so much experience in shipping that his father was able to retire and leave the business to Samuel, thereby resulting in a name change from A. Cunard and Company to S. Cunard and Company. A year later, in 1815, Samuel pulled off a major coup by negotiating the first of numerous contracts with the British Government to carry mail between Great Britain, Halifax, Saint John's, and Boston, as well as to Quebec City, Montreal, and along the St. Lawrence River as a result of an ingenious plan Samuel worked out with the Overland Express. This coup provided a steady source of income for the Cunard company for decades, as well as the financial stability, security, income, and incentive needed to build a powerful shipping empire.

Samuel Cunard was a remarkable businessman and entrepreneur. By 1825, he had acquired more than 30 ships and was carrying mail, passengers, timber, dried fish, rum, sugar, sheep, coffee, and flour on a regular basis between Great Britain, the Caribbean, South America, Newfoundland, and Canada (or British North America as it was called at that time). As John Boileau points out in his book *Samuel Cunard: Nova Scotia's Master of the North Atlantic*, "The worldwide revolution in transportation and communication that Samuel Cunard started in 1840 continues to this day. The ideas and the legacy of the little man from Halifax who became one of the greatest entrepreneurs of his time have come a long, long way."[2]

By 1848, the Cunard Line had nine steamships sailing the Atlantic Ocean on a weekly basis in both directions every month of the year. In that year, Samuel moved to England and began to run the business out of London rather than Halifax. The Cunard

---

2. John Boileau, *Samuel Cunard: Nova Scotia's Master of the North Atlantic* (Halifax: Formac Publishing Company, 2006), p. 6.

Line built its reputation on speed, comfort, and the ability to provide services on a regular, sustained, and systematic basis. As John Boileau has also noted, "Cunard's genius led to a major paradigm shift: the first use of regularized transportation and communication by steamship service with advertised and standardized sailings, the nautical equivalent of a railroad timetable."[3] But even more was required. The Cunard Line also built its reputation on safety, often boasting that "it never lost a passenger or a letter." As Mark Twain said following a trip across the Atlantic on a Cunard ship, "it is rather safer to be on board their vessels than on shore."

With outstanding achievements and an excellent reputation like this, the Cunard Line was reorganized internationally in the latter part of the nineteenth century and early part of the twentieth century and began to assert itself in earnest as *the* major shipping company in the world. In 1878, it pioneered a popular 13-week cruise in the Mediterranean and Black Seas that gave rise to the modern cruise holiday that is very popular with tourists and travellers today. Also included among its many "firsts" around that time were the *Servia*, built in 1881 and the first steel ship to be built by the Cunard Line and the first ship in the world to use electric lights; the *Campania*, built in 1893 and the first twin-screw mail ship built without sails and using steam turbines; the *Lucania*, built in 1901, which set a record of five and a half days in each direction across the Atlantic that year and was the first ship to be equipped with wireless from which Marconi conducted many of his wireless telegraph experiments; the *Lusitania* and the *Mauretania*, built in 1907 and unrivalled in the history of steamship navigation for their speed, elegance, size, beauty, durability, and money-making

---

3. John Boileau, *Samuel Cunard*, p. 5.

ability; and, somewhat later, the *Queen Mary* and the *Queen Elizabeth*, which were constructed in the 1930s and remain well known even today although they are no longer in service. The two *Queens* carried more than a million and a half soldiers and large amounts of supplies to various destinations during and after the Second World War, as well as millions of passengers before and after the war.

This all emanated from the vision, determination, and creativity of Samuel Cunard, who was born and spent most of his life—and certainly virtually all of his working life—in Halifax. Cunard was one of the greatest visionaries and entrepreneurs Canada has ever produced. As a tribute to him, there is a marvellous statue of this old "steam lion," as he was often called, located on the pier in Halifax. Cunard is gazing out on Halifax harbour, but beyond that to the Atlantic Ocean, England, Europe, and indeed the entire world. Following a brief summary of his incredible accomplishments, there is an inscription on the base of the statue stating in large, bold letters *Samuel Cunard. Haligonian. World Benefactor*. Cunard is described as "a visionary who foresaw steam power replacing sail power on the North Atlantic." The summary goes on to say that "the advent of steam on the North Atlantic forever altered commence and communication between the Old and New Worlds." Clearly a fitting tribute to the man who transformed water transportation and shipping in the Atlantic Ocean and eventually the entire world.

Cunard was not the only Maritimer to establish an international reputation in water transportation. Several others also made their mark on the world by helping to change the face of water transportation in the nineteenth century. Two of the most notable were John Patch and Robert Foulis.

Like many water transportation vehicles and devices, the screw propeller can be traced back to ancient times. The Greek inventor Archimedes, for example, proved the power of a helical screw to move water as far back as 300 BCE. Nevertheless, numerous centuries were to pass before there was any concrete evidence that a propeller could be used effectively to power boats, despite the fact that there was a great deal of interest in this and many drawings were made of proposed devices capable of doing so over the centuries. Some claim that John Stevens of the United States was the first person to create a workable screw propeller since steam screw propellers were used on boats on the Hudson River between 1802 and 1806. Others claim Josef Ressel of Czechoslovakia created a screw propeller with multiple blades fastened around a conical base and tested this propeller in February 1826 on a small ship. He received an Austro-Hungarian patent for this in 1827, and was successful in using a bronze screw propeller on an adapted steamboat in 1829. However, there is a great deal of empirical evidence to confirm that John Patch of Nova Scotia played the most important role of all in the creation and development of a viable screw propeller.

Patch experimented with screw propellers in the late 1820s and early 1830s. The significance of his achievements in this area should not be underestimated or ignored. The screw propeller made possible the transition from transportation by wind and sail to steamships and eventually modern ocean liners and freighters. Ironically, it also laid the foundation for another Canadian to made a similar creative contribution in another area of transportation to be discussed momentarily.

Unlike John Patch, who was born in Canada, Robert Foulis, another highly creative and prolific Maritimer, was born in another country. In this case, it was Scotland, like many other famous Canadian creators in earlier periods of Canadian history.

Foulis came to Canada at a very early age and settled in St. John, New Brunswick in 1818 where he became actively engaged in many creative endeavours as a result of his background in engineering and mechanics. However, his crowning achievement was the invention of the steam fog horn or "steam whistle." He presented his plan for this invention to the Lighthouse Commissioners of New Brunswick in 1853. This eventually led to the installation of the first steam fog horn in the world at Partridge Island, New Brunswick in 1860. This invention did a great deal to improve safety on the Atlantic Ocean and eventually the entire world, largely by transmitting automatically-coded blasts of sound that could be heard several miles away to warn ships of impending danger due to their proximity to land and other vessels. As one sea captain put it, "the man who invented that fog whistle should get to heaven if anyone does."

A similar device—the diaphone—was also invented in Canada, but this time by a person from Ontario. This sound-making device was used around the world as a fog horn in the first half of the twentieth century because it produced a much deeper and more powerful sound than the steam fog horn designed by Foulis. It also projected this sound over substantially longer distances and larger areas. The individual who invented it, John Pell Northey, worked at the University of Toronto and based his design on the organ stop invented by Robert Hope-Jones, creator of the Wurlitzer organ. Northey added a secondary air supply to power the piston on its forward and backward motions, giving rise to a much stronger sound.

In 1903, Northey began manufacturing the diaphone

commercially at his Diaphone Signal Company in Toronto. The company produced a wide range of diaphones, including the "Type F" version that produced a tone of about 250 hertz. It was immediately utilized by most lighthouses and ships around the world. Not content with this, Northey's son Rodney redesigned the Type F version to sustain a second low tone, thereby creating the familiar two-tone signal that was used in most lighthouses and boats in Canada, the United States, and other parts of the world for more than half a century. Most of these diaphones were removed or disconnected when lighthouses were automated in the 1960s and 1970s, although some still remain and are in good working condition today.

About the same time that the diaphone was being manufactured in Canada and sold throughout the world, another inventor whose origins can be traced back to Scotland—Alexander Graham Bell—was involved in the construction of a hydrofoil boat, as noted earlier. This boat includes a wing-like structure that is mounted on struts beneath the hull so that the boat travels seemingly on top of the water. The struts are designed to lift the boat partially out of the water in order to reduce the amount of drag.

Although the hydrofoil was not invented by Bell, he made many seminal contributions to its evolution and use at his summer residence and research facilities in Nova Scotia—contributions that were instrumental in assuring a prominent place for Canada in the development of the hydrofoil for many years as well as its use throughout the world. Here is how this came about.

While a number of inventors from England and the United States were experimenting with hydrofoils in the late nineteenth and early twentieth centuries, the first really successful hydrofoil boat was created by Enrico Forlanini of Italy. It reached speeds of over 70 kilometres per hour on Lake Maggiori between 1905 and 1911. However, it is clear in retrospect that two of the greatest pioneers in the evolution of the hydrofoil were Alexander Graham Bell and Casey Baldwin together with Forlanini. Bell and Baldwin began experimenting with the hydrofoil boat in the summer of 1908. In 1910–1911, they met Forlanini in Italy, where they rode in his hydrofoil boat on Lake Maggiore. Returning to Bell's laboratory at Beinn Bhreagh on the Bras d'Or lakes in Cape Breton, Bell and Baldwin experimented with a number of designs that culminated in the construction of the HD-4 in 1917. It was the fastest boat in the world for a time, providing a smooth ride, taking the waves without too much difficulty, and accelerating very rapidly. On September 9, 1919, it set a world marine speed record of 114.04 kilometres per hour—a record that stood for more than a decade.

While Bell died in Nova Scotia three years later, Baldwin continued to work on designs for more effective hydrofoil boats, including an experimental smoke-laying hydrofoil during the Second World War. In 1952, the Canadian Armed Forces built and tested a number of hydrofoil boats successfully, including two vessels called *Bras d'Or*. They also built and tested a high-speed anti-submarine hydrofoil, the HMCS *Bras d'Or*, in the late sixties. It was deemed to be the fastest warship in the world, but the program that produced it was cancelled by the Canadian military in the early 1970s due to the shift away from anti-submarine technology and warfare.[4] Many others countries were also involved in the development of hydrofoils in the years to follow, especially Russia where they played and continue to play a key role in transportation. However, there

---

4. John Boileau. *Fastest in the World: The Saga of Canada's Revolutionary Hydrofoils* (Halifax: Formac Publishing Company, 2004).

is no doubt that Canada's contributions in this area paved the way for a number of important global developments, as well as assisting in the evolution of the hovercraft with which the hydrofoil shares certain similarities as a result to its ability to travel on and above water rather than in it.

This is not the only area where Canadians have made creative contributions that have had an important impact on transportation in the world. When attention shifted from water transportation to rail transportation in the middle and latter part of the nineteenth century, Canadians were once again in the forefront of important developments in this area with international implications and consequences. Spurred on by the industrial revolution, water transportation was rapidly giving way to rail transportation as one of the most effective and lucrative ways of moving people, products, and resources over long distances and of creating prosperity and wealth.

Canada was in an ideal position to capitalize on these developments. The country's vast expanse made visions of rail transportation exciting, as railways were seen by many as the key to linking communities, towns, cities, and huge geographical areas together and facilitating greater agricultural and industrial production. As a result, Canada shifted rather rapidly from the construction of boats and ships to the creation of railroads and trains. Doing so made a lot of sense for a country where waterways and ports were frozen over for a significant part of the year and which had an inadequate system of roads. The federal government played a major role in the development of railroads, not only for economic and financial but also for political reasons. It was anxious to link the country together at a time when there was an expanding interest in doing so across Canada, as well as countering the ongoing threat that the provinces in existence at that time would go their separate ways or get swallowed up in the American "Manifest Destiny" movement.

Developments like these, and especially Canada's enthusiastic embrace of transportation by rail, led to many creative achievements in the development of railroads and rail technology in the country in the following decades. This included the creation of the sleeper car by Samuel Sharp in 1857; development of an air-conditioned coach by Henry Ruttan in 1858; and the invention of a locomotive braking system by W. A. Robinson in 1868. Then, in 1869, a dentist named J. E. Elliott created a "rotary snow plough." It was refined over subsequent decades to facilitate rail transportation during the winter months in Canada, the United States, and other parts of the world.

As important as this technology was, it was the construction of the Canadian Pacific Railway (CPR) in the latter part of the nineteenth century that was one of Canada's—and the world's—greatest and most creative achievements in the railroad era. The CPR was completed in 1885 when the laying of an incredible "band of steel" across the gigantic land culminated with the hammering in of "the last spike" at Craigellachie in Eagle Pass, British Columbia. What was particularly creative about this achievement was not so much the construction of the railroad itself, since railroads had of course existed for some time prior to this and were increasingly common throughout the world, but rather the scale of this remarkable accomplishment and the ingenuity, determination, and imagination needed to plan and realize it. This included overcoming numerous technical, financial, and administrative problems and driving the railway through some of the most difficult and treacherous terrain in the world in northern Ontario, western Alberta, and eastern British Columbia. Built primarily with British, Canadian, and American capital and expertise as well as Chinese immigrants and labour, the CPR stands out as one of the greatest engineering feats in world transportation history.

While the observation to follow is based on conjecture on my part because there is no factual evidence to back it up, nonetheless it is very possible that the construction of the CPR provided the impetus and motivation that were needed to build national, intercontinental, or transcontinental railroads in other countries of the world, especially in view of Canada's colossal size and small population. There was probably nothing quite like a prototype of this type to stimulate similar developments in Russia, India, China, and other parts in the world.

Just as water transportation gave way to rail transportation at a certain time in Canadian and world history, so in turn rail transportation gave way to air transportation. This occurred in the twentieth century, and, once again, Canadians were in the forefront of some of the most creative and essential developments in this area.

In 1909, as mentioned earlier, J. A. Douglas McCurdy piloted the first powered flight in the British Empire when he flew the *Silver Dart* at Baddeck, Cape Breton. He worked on this aircraft with Alexander Graham Bell, F. W. "Casey" Baldwin, and especially Bell's wife Mabel, who was also a very prominent person in the early development of the aviation industry through the Aerial Experiment Association she created at Baddeck. McCurdy was also the first person to fly an airplane out of sight of land, in this case off the coast of Florida. Canadians were back in the news in this area once again a hundred years later when Todd Reichert flew a human-powered ornithopter, an aircraft with bird-like wings called the *Snowbird*, over 145 metres in 2009, thereby fulfilling a dream that can be traced back to the myths of ancient Greece.

But the most important creative contribution by a Canadian to air transportation with world implications and consequences is undoubtedly the creation of the variable pitch propeller by Wallace Rupert Turnbull. Turnbull was born in Rothesay, New Brunswick in 1870 and graduated from Cornell University in Ithaca, New York in engineering before going to Berlin and Heidelberg to undertake post-graduate studies. Like many of Canada's creative people, Turnbull worked in other parts of the world during important stages of his life, most notably in England during the First World War where he developed the first working model of the variable pitch propeller in 1916, as well as in Harrison, New Jersey, where he worked for a time in Thomas Edison's lamp factory. But the bulk of his work was conducted in a barn on his property in Rothesay focusing on his two passions: propulsion and aerodynamics. He built Canada's first wind tunnel in his workshop, and delved deeply into bomb sights, torpedo screens, hydroplanes, and, towards the end of his life, capitalizing on the power of the tides of the Bay of Fundy.

Turnbull's variable pitch propeller wasn't patented until 1922 and was then successfully tested in 1927 at Camp Borden air base in Ontario. Just as the creation of the screw propeller by Joseph Patch and others had a powerful impact on water transportation throughout the world, so the creation of the variable pitch propeller by Wallace Rupert Turnbull had an equally powerful impact on air transportation. By changing the angle or pitch of the propeller blades, the variable pitch propeller made it possible for aircraft to take off easier, climb higher, fly faster, establish cruise control more readily, and improve fuel efficiency. All these factors made it possible to carry much heavier loads and many more passengers than aircraft were able to do prior to this time, thereby playing a key role in the development of commercial aviation throughout the world. News of this remarkable invention spread rapidly in the aviation field. When the Canadian government refused to get involved in this timely and important invention, Turnbull sold the patent rights to the variable pitch propeller to Curtiss Wright Corporation in the United States and the British Aeroplane Company in England. Since these were the two largest producers of airplanes in the world, Turnbull's propellers were eventually sold and utilized worldwide, impacting virtually every country in the world.

The variable pitch propeller is seen by many aviation experts as one of the most valuable contributions to aeronautics, making commercial aviation feasible and profitable in financial terms, and therefore a reality. In doing so, it reduced the dependence of the commercial aviation industry on public sector grants and government subsidies, thereby helping to fuel the rapid expansion of commercial and transport aviation throughout the private sector in most parts of the world. Given the size of this industry today and its phenomenal expansion over the last 50 years—both in terms of cargo and passengers—there is no doubt that the variable pitch propeller created by Turnbull was a remarkable contribution to the whole world.

This by no means completes the story as far as Canadian creativity in aviation and its impact on the world is concerned. Canadians have also made seminal contributions to the development of many other aspects of the aircraft industry and aviation technology, such as the Norseman Bush Plane created by Bob Noorduyn in 1935. It helped immensely to facilitate exploration, discovery, and search and rescue in Canada's far north and other remote areas of the world. Moreover, Canadians have also been pioneers in the development of aircraft capable of taking off and landing on short runways, or what are called "STOL" (short take-off and landing) aircraft. The first STOL aircraft, the *Beaver*, was created at de Havilland Aircraft of Canada in 1947. The United States Armed Forces ordered more than 900 of these planes, described as "the best small utility aircraft in the world," and operators in sixty-two other countries also made purchases. The planes operated in the Middle East, high in the Andes in South America, and in the polar regions of the world. They also operated in Korea and Vietnam, dropping supplies and evacuating casualties. This opened the doors for the creation of many other STOL aircraft, including the *Turbo-Beaver, Otter, Twin Otter, Caribou,* and *Buffalo* that have increasingly been in demand throughout the world. Canada has also played an important role in the creation of other small aircraft ideally suited for urban and intra-city travel, such as the *Dash-7* and *Dash-8* that were designed and built at De Havilland Aircraft in Toronto.

Although Canada has made many creative contributions to the different types and modes of transportation in the world, none has been more distinct or unique than the creation of vehicles that travel over snow, such as the snowmobile and Ski-doo. They make a great deal of sense for countries covered in snow for a significant part of the year.

The person who invented these popular devices, Joseph-Armand Bombardier, was born in Valcourt, Quebec in 1907. He dreamed of creating vehicles that could glide over the snow in his childhood and youth. Canadians had been using vehicles that could travel over snow for many centuries prior to this, but they were largely sleighs and toboggans. Bombardier had something different in mind—*motorized* vehicles that could travel over the snow. He was no doubt inspired by the fact that the snowfalls and snowdrifts in Quebec as noted earlier were so relentless and deep at times that many roads were closed or blocked during the winter months, thereby preventing the use of cars, trucks, and other vehicles.

Bombardier realized his vision in 1922 when he took an old Ford his father had given him, removed the motor, and attached it to the frame of a four-passenger sleigh. He then installed a huge wooden airplane propeller to the drive shaft and produced the first *power-driven* mechanical vehicle—a snowmobile—capable of travelling over the snow. Like many other famous inventors, Bombardier spent the bulk of his life after this perfecting this invention and creating many others like it. In so doing, he set in motion a chain of events that has had an important impact on Canada and the world since that time. After being granted a patent for his snowmobile in 1937, Bombardier immediately posted a sign above his garage—L'Auto-Neige Bombardier—and started manufacturing snowmobiles commercially. It proved to be a great success.

In 1959, Bombardier invented the Ski-Doo, a snowmobile intended primarily for personal, family, and recreational use that took Canada and the world by storm. Not only did it sell well in all parts of Canada, adding a whole new dimension to the meaning of winter, recreation, and fun in the snow, but also it resulted in the creation of a huge industry. The Ski-Doo sold equally well in countries that experienced winter conditions similar to those in Canada, such as Sweden, Norway, Denmark, Greenland, Iceland, and Russia. By 2000, more than two million Ski-Doos were being manufactured and sold worldwide on an annual basis. Bombardier's company quickly became one of the largest manufacturers of winter transportation vehicles in general—and winter recreational vehicles in particular—in the world.

Regrettably, the company later fell on hard times and experienced financial and administrative problems as well as some very real technical difficulties, causing it to spin off the division that produces and sells Ski-Doos and other types of recreational equipment as a separate company, BRP, still based in Valcourt. The former parent company continues to play a major role in global transportation, largely through the manufacture of subway and street cars, rail cars, and business jets (the regional jet business was sold to Japan's Mitsubishi in 2019). In 2017, it had 73 production and engineering sites in 28 countries around the world, with 66,000 employees worldwide and reported a net income for 2016 of US$553 million—a substantial contribution to the world from a country with a relatively small population.

Bombardier's achievements were accompanied by another creative Canadian innovation in winter transportation that has proven successful throughout the world. This is the invention of the snowblower by Arthur Sicard in 1925. It consisted of three parts: a four wheel drive truck chassis and truck motor; a "snow scooping" section; and a snow blower with two adjustable chutes and a separate motor. This device enables drivers to throw snow over 90 feet away, or directly into the back of dump trucks, and worked equally well in hard, soft, and packed snow. Sicard's invention played an important role in making it possible for people in Canada and other parts of the world where snow and winter conditions prevail to come to grips with large accumulations of "the white stuff" on roads, highways, driveways, and airport runways. As a result, the Sicard Company is now exporting snow blowers to many countries in the world where snow prevails.

These contributions to transportation over snow have been complemented by the creation of other motorized vehicles and devices by Canadians capable of facilitating transportation over other types of difficult terrain. Most notable in this regard are the *Giraffe*, a wheeled vehicle raised and lowered by hydraulic power; the *Ridgewater*, a four-vehicle device capable of carrying heavy loads over rough, tough, and treacherous terrains; and the *Rat* and *Jigger*, vehicles that travel equally well over land, water, snow, muskeg, and swamps. They are now used in many parts of the world for these purposes.

One would expect a country like Canada that has been engaged in many creative developments in transportation by water, land, air, over snow, and in rough terrains to be engaged in similar developments in space as well. Canada does not disappoint in this area. While the country's involvement in space started somewhat later than the two leaders and pioneers in the field—Russia and the United States—Canada made this a triumvirate in some ways when it got involved in space exploration, transportation, and technology as a founding member of COSPAR—the Committee on Space Research—of the International Council of Scientific Unions in 1958.

Since that time, Canada has distinguished itself throughout the world as a result of numerous major initiatives in space exploration. Included among these initiatives are the launching of Alouette 1 in 1962 to study the ionosphere, making Canada the third country in the world with a satellite in space; the creation of Spar Aerospace Limited at de Havilland in Toronto; the launching of Anik A1, the first domestic (as opposed to international) communications satellite in geostationary orbit; and the sending of several astronauts into space, including Marc Garneau, Roberta Bondar, Chris Hadfield, Julie Payette, Canada's current Governor-General, and, most recently, David Saint-Jacques.

The crowning achievements in this area are undoubtedly the *Canadarm* and the achievements of Chris Hadfield. The origins of the *Canadarm* can be traced back to 1974 when Canada was awarded a contract from NASA to design, develop, and build a Shuttle Remote Manipulator System—or SRMS for short—for the space shuttle. It worked so well that it instantly gave Canada a superb international reputation in robotics innovation and space technology. With copper wiring for nerves, graphite fibre for bones, and electric motors for muscles, the *Canadarm* acts very much like the human arm with two rotating joints at the shoulder, one at the elbow, and three at the wrist, along with a "hand." Constructed by Spar Aerospace Limited at its plant in Baie d'Urfe, Quebec, the *Canadarm* was used in many space shuttle missions due to the fact that it weighed less than 480 kilograms, was only 15 metres in size, and capable of lifting from 30,000 to 266,000 kilograms in the weightlessness of space. After the end of the Space Shuttle program, the *Canadarm* was retired, but its successor, the *Canadarm2*, is now used on the International Space Station. Whereas the elbow rotation of the first *Canadarm* was limited to 160 degrees, the rotation of *Canadarm2* is 540 degrees since each of the joints is able to rotate 270 degrees in either direction.

Also worth discussing in detail are Chris Hatfield and his creative contributions to space exploration. Born in 1959, Hadfield was raised in a corn-producing area of Ontario and became interested in flying and becoming an astronaut at a very early age after he watched the Apollo 11 moon landing on TV.

After receiving military and flying training in the Canadian armed forces and taking part in an exchange program with the United States Air Force, Hadfield was accepted into the Canadian astronaut program at the Canadian Space Agency in 1992. Following this, he acquired a great deal of expertise and experience as an astronaut in various locations throughout the world, including walking in space and helping to install the *Canadarm 2,* and was eventually made commander of the International Space Station as part of Expedition 35 in 2013. In this capacity, he was responsible for a crew of five astronauts and gained an international reputation and a huge global following for depicting what it is like to live in space with humour, imagination, and vitality, as well as singing songs and making music. In so doing, he once again made space travel an area of interest for millions of people throughout the world, especially young people who are now thinking seriously about careers as astronauts, as well as David Saint-Jacques, a fellow Canadian, who followed in his footsteps and travelled in space in 2019,

With so many creative achievements in transportation, it is no coincidence that Canadians have also made many creative contributions to map-making and cartography. These contributions are even more understandable in view of the colossal size of the country and the fact that early in Canada's history countries like France, Great Britain, and other European powers were engaged in a fierce battle to stake out claims to huge tracts of land throughout the North American continent, in order to determine what natural resources existed here and how these resources could be capitalized on.

Important contributions to this field were made by Samuel de

Champlain, David Thompson, and especially Samuel Holland, who is credited with setting the stage for Canada and Canadians to become world leaders in cartography and map-making when he was appointed Surveyor-General of Canada following the signing of the Treaty of Paris in 1763. These contributions include mapping a considerable part of what eventually became Canada; the creation of the second national atlas in the world after Finland—*The Atlas of Canada*—by the Department of the Interior in 1906; major contributions to undersea and shoreline mapping; numerous firsts in magnetic surveying when Sir Edward Sabine of Toronto was involved in creating the first apparatus for detecting the changing magnetic fields of the earth in Madras, India, Melbourne, Australia, Saint Helena Island, and Toronto; and the contributions made by Sir William Logan, founder and first director of the Geological Survey of Canada, on which the country's mining industry depended for so many years and still depends today.

This early leadership was sustained throughout the twentieth century when Dr. Les Howlett developed the camera calibrator at the National Research Council in Ottawa in 1945 that set the standard for cameras used in aerial photography; authorities at the federal Department of Forestry and Department of Mines made many original contributions to cartography and map-making; people like J. M. (Monty) Bridgman, T. J. Blachut, Stanley Collins, Douglas N. Kendall, Gilbert L. Hobrough, and many others created numerous devices capable of mapping the lands, forests, and mines of Canada; and Roger Tomlinson made a highly inventive contribution to visualizing data on maps that is playing a major role in urban mapping today. But the greatest and most creative Canadian contribution in this area came from U.V. Helava. He set the international map-making community on fire by creating the analytic plotter in 1957, making it possible to take photographs from orbiting satellites. In so doing, he helped maintain Canada's reputation as a global leader in map-making, a reputation that is still intact today.

With numerous "firsts" in such areas as the camera calibrator, the radar profile recorder, the analytic plotter, stereo-orthophotography mapping techniques, the automatic image correlator, the Gestalt photo mapper, geographic information systems, and others—many of which were developed at the National Research Council in Ottawa or by researchers and inventors connected with the Council—it is understandable why it is that over sixty percent of the map-making technology used throughout the world today is based on Canadian technologies and inventions.

If transportation is to function effectively, it is not only necessary to deal with a host of difficult spatial problems concerned with moving people, products, and resources over long and short distances. It is also necessary to come to grips with a series of complex temporal problems related to keeping time and the development of an accurate, authentic, and effective time-keeping system. The one is as essential as the other.

As was discovered in the development of continental rail systems, it is very inconvenient and expensive for different countries and regions to use different railway gauges, since rail cars then have to be unloaded and reloaded at border crossings and check points. The same holds true for the keeping and recording of time. Without a uniform system of time, transportation can easily be disrupted and delayed. This was a particularly pressing problem for Canada as the second largest country in the world, one stretching across a quarter of the world's 24 time zones.

Just as some of the most important transportation inventors and inventions in the world have been Canadian or had close affiliations with Canada, so one of the greatest inventors in the world with respect to time was also a Canadian. It was Sir Sandford Fleming. He developed a great deal of interest in time from the problems he encountered living and working in Canada, largely as a result of the colossal size of the country and its complex spatial and temporal requirements and challenges.

As every Canadian learns in school, Fleming played a crucial role in the creation and utilization of *standard time* throughout the world. The world owes this creative genius a great debt for the pioneering role he played in this matter.

Fleming immigrated to Canada from Scotland when he was very young. He went on to become a skilled engineer and surveyor as well as a distinguished inventor and scientist. Not only did he design Canada's first postage stamp, but also he left behind a large body of survey information and map-making equipment when he died, was a founding member of the Royal Society of Canada, and was the founder of the Royal Canadian Institute. He also played a seminal role in the construction of the Intercontinental Railway and especially the Canadian Pacific Railway. It is generally agreed that the CPR would not have been built without Fleming's vision, determination, creativity, boundless energy, and enthusiasm.

Fleming applied these same qualities to the creation and development of a universal system of time. The impetus for this occurred when Fleming was delayed in a train station in Ireland for more than twelve hours because of a typographical error that mixed up 5:35 *a.m.* and 5:35 *p.m.* This caused him to think seriously about time and how it could be kept most effectively. He then spent the next few years of his life studying the diverse time systems used in different parts of the world.

Following this, he plunged into making the case for an international time system that would be applicable and acceptable to all countries. After experiencing numerous setbacks, dead ends, and rebuffs, Fleming finally succeeded in convincing a number of major scientific authorities and institutions in different parts of the world that his plan for a universal time system made a great deal of sense. This eventually led to the adoption of an international time system at the International Meridian Conference in Washington in 1884, a conference that Fleming was instrumental in organizing and that was based on his ideas. This system quickly became known throughout the world as the *Greenwich Mean Time System,* since time zones were marked off as one moved eastward or westward from Greenwich, England (England at that time was home to two-thirds of the world's shipping companies). Despite Fleming's success in promoting his ideas, it was not until 1929 that most major countries in the world adopted a system of time zones defined by the difference between local time and Greenwich Mean Time.

While it took many decades for Fleming's international time system to fully become a reality, his creative ideas in this area were eventually acknowledged and provided the foundation for the present system of time zones used everywhere in the world today. Life is much easier and more convenient and predictable for all people and all countries as a result of this, as the Fleming system of time has yielded countless economic, financial, transportation, communications, recreational, tourist, and travel benefits in all parts of the world.

If Canada's creativity in transportation and developing a suitable time-zone system has had a major impact on the world, so has its creativity in communications. Here, as well, much of the creative activity in this area took place initially in the Maritimes, due largely to the pressing need to connect the old world with the new.

By the 1840s, an American painter and inventor named Samuel Morse had created a single-wire telegraph and a way of sending messages on it—Morse code—for which he is rightly famous. He also felt it was necessary and possible to lay a cable across the Atlantic Ocean that would facilitate communication between Europe and North America by telegraph. By 1850, Bishop John T. Mullock, head of the Roman Catholic Church in Newfoundland (which became a province of Canada not quite one hundred years later in 1949) was advocating construction of a telegraph line that would run from St. John's to Cape Ray, as well as a cable across the Gulf of St. Lawrence from Cape Ray to Nova Scotia along the Cabot Trail.

Ideas like these resonated strongly with Frederick N. Gisborne, an engineer and communications expert who had emigrated to what was soon to become Canada from England. Along with several others—especially Cyrus West Field, a highly successful New York businessman and entrepreneur—Gisborne played a seminal role in laying the first undersea telegraph cable in North America. This occurred in 1852, when Gisborne and his colleagues laid a cable from Cape Tormentine, New Brunswick to Carleton, Prince Edward Island that was impervious to salt-water corrosion. Following this, Gisborne, Field, and others established the New York, Newfoundland, and London Telegraph Company for the purpose of laying a cable across the Atlantic Ocean. After many tries and failures due to the splitting of cables and other technical difficulties, a cable was finally laid across the Atlantic Ocean from Trinity Bay in Newfoundland to Valentia in Ireland. This event occurred in 1858 and made it possible to transmit the first telegraphic message from North America to Britain on August 16, 1858. Queen Victoria responded to this message by sending a reply to President James Buchanan in the United States. The Queen's message read: "Europe and America are united by telegraph. Glory to God in the Highest, on earth peace, good will towards men." It took 16 hours to transmit this message.

Not surprisingly, Canadians were also involved in laying the first marine cable across the Pacific Ocean almost 50 years later, thereby complementing the feat achieved by West, Gisborne, and others in 1858. This occurred in 1902, when a cable was laid from Vancouver to Southport, Queensland in Australia. After centuries of relative isolation, the world was starting to become "all of a piece" in a way it never had been before. While this process took time, the invention to be discussed next played a crucial role in making all this possible.

Of all the devices created by human beings, few have been as monumental or have had as profound and lasting an effect on the world as the telephone. The telephone was an ideal device for Canada and Canadians because it provided communication over short and long distances at a time when the country's citizens were extremely dispersed and isolated from one another. Since many Canadians worked on farms and in rural areas at this time, they had little or no contact with family members, friends, and relatives living in other parts of the country. The telephone changed this, eventually connecting the entire country and the whole world through the transmission of the human voice over wires rather than in person.

This remarkable device was created by Alexander Graham Bell in 1875. While Canadians were quick to claim the telephone as a Canadian invention and Bell as a Canadian, Americans were equally quick to claim the telephone as an American invention and Bell as an American. This was because Bell conducted much of his pioneering work on the telephone in both Canada and the United States. Not only did he spend a great deal of time in Boston, where he taught at a school for the deaf and his wife Mabel's family was located, but also he spent an enormous amount of time in Canada, first in Brantford, Ontario where his own family lived after they came to Canada from Scotland and Bell did much of his early and most original work on the telephone, and later in Cape Breton, where he and Mabel had their summer residence and research facilities as mentioned earlier. While Bell became an American citizen ten years after the telephone was invented, he remained very attached to Canada, as is well documented in the Bell Museum in Nova Scotia.

The dispute over whether the telephone was a Canadian or American invention was put to rest by Bell himself when he said the telephone was "invented" in Brantford, Ontario in 1874 and "born" in Boston, Massachusetts in 1875. It is not just this fact, however, that confirms that the telephone was more a Canadian invention than an American one. Canada was also the place where the first long distance telephone call in the world took place. It occurred from Mount Pleasant, Ontario to Brantford, Ontario on August 3, 1876. A second and longer call was made from Brantford, Ontario to Paris, Ontario on August 10 of that same year. The latter call was achieved by using power from a battery located in Toronto, some 70 miles away.

Much like Thomas Edison, Bell is viewed by many as one of the greatest inventors of all time, largely because of the numerous inventions he created and the remarkable communications and social transformations they produced. As far as the telephone is concerned, this is due not only to the prolific work Bell did as the inventor of the telephone, but also his work as a speech therapist for the deaf, the scientific approach he took to the harmonic telegraph, and the invention of a successful microphone that paved the way for the creation of the telephone. The impetus for his work in this area probably came from the fact that both Bell's mother and wife were deaf, something which goes a long way towards explaining his consuming interest in deaf people, deafness, the telephone, and other speech and acoustical devices.

While other inventors such as Elisha Gray in the United States were conducting similar experiments at about the same time, Bell is generally recognized as the principal inventor of the telephone because he received the patent for this—U.S. Patent number 174,465, which has often been called the most important patent ever issued in the United States—on March 7, 1876. And this is not all. Bell was more aware of the commercial implications, possibilities, and benefits of the telephone than his competitors and therefore actively engaged in its commercial development to a much greater extent, primarily by participating in the creation of the Bell Telephone Company in the United States in 1876–77, a development that eventually led to the creation of the Bell Telephone Company of Canada.

Like the automobile and many other inventions, the importance of the telephone was immediately recognized by many people who wanted a telephone of their own. As a result, the use of the telephone throughout North America and the world expanded rapidly. By March 1880, there were 138 exchanges in operation in the United States alone with more than 30,000 subscribers. By 1887, this number had ballooned to 150,000 in the United States, 12,000 in Canada, 26,000 in the United Kingdom, 22,000 in Germany, 9,000 in France, and 7,000 in Russia. By 1904, over three million telephones were connected by manual switchboard exchanges in the U.S. alone.

Commercial expansion of telephone use throughout the world was slowed only by a number of technical problems that had first to be overcome. These included the poor quality of the sound, competition between various telephone companies, the nature and effectiveness of the receivers, the development of switchboards and telephone exchanges, and the fact that many towns had multiple terminals rather than a single terminal. Even more problems were encountered in the international expansion of the telephone, such as the type of wires used in long distance transmissions, the laying and utilization of various types of cables and coils, the boosting and amplification of the signals, and interference from other technological devices and sources.

Nevertheless, these problems were eventually overcome and the telephone quickly became one of the most popular, frequently used, and most successful communications devices ever created. By 1915, the first coast-to-coast telephone call had been made in the United States from New York to San Francisco. Then, in 1926, two-way voice communication was achieved across the Atlantic Ocean by means of radio transmission. This was followed by the laying of the first transatlantic submarine cables for telephone use in 1955, and the launch of the first international communications satellite, Telstar, in 1962. These developments were later followed by a major shift from copper wires to fibre optics and the introduction of the Internet and other contemporary forms of communication.

This by no means completes the story as far as the

international diffusion and use of the telephone is concerned. Whereas the telephone had depended on transmission of the human voice over wires during most of this time, another creative breakthrough was achieved in this area when the world's first mobile phones, or "cellphones," were created. This produced yet another revolution in communications, one that has been much more rapid than the revolution brought about by the original invention and development of the telephone. The number of mobile phone users worldwide was expected to surpass the five billion mark in 2019. It boggles the mind to think of how this invention by Alexander Graham Bell in 1875 has impacted on the entire whole world since it was first created. Not only is the world totally connected by many different types of phones today, but also the amount of time and money spent on this form of communication has been astronomical, with no end in sight.

Canada was a hot bed of creativity in communications when the telephone was invented and for some time thereafter. This is particularly true for long-distance communication by radio, due largely to the creative genius of Guglielmo Marconi. This world-famous inventor, electrical engineer, and entrepreneur was born in Italy, spent much of his time in England, and was interested in creating a radio telegraph system that was capable of transmitting signals by radio waves rather than wires in order to compete with and complement the system of transatlantic telegraph cables that existed at that time.

After years of experimentation, Marconi was finally successful in transmitting a signal from Poldhu in Cornwall, England to Signal Hill in Newfoundland some 3,500 kilometres away on December 12, 1901. Since there was some controversy over this achievement because there was no official confirmation that the signal was actually received, Marconi followed this up by transmitting a signal in the opposite direction from Glace Bay, Nova Scotia to England on December 17, 1902. This was the first *confirmed* radio signal to cross the Atlantic Ocean from North America to Europe. By this time, Marconi had outfitted two of the Cunard Lines' most important ships—the *Lucania* and the *Campania*—with a wireless system that made it possible for these ships to send and receive messages in this form. By 1903, these ships were sending countless messages back and forth, including an onboard *Cunard Daily Bulletin* that quickly became a regular feature and very popular with passengers and crew.

It was achievements like this that account for the fact that Marconi was awarded a Nobel Prize in Physics in 1909 along with Karl Braun of Germany for their contributions to what was called "wireless telegraphy" at that time and is now called "radio." Marconi's achievements yielded other benefits as well. When the *Titanic* sank in 1912, a Cunard ship—the *Carpathia*—saved many of the shipwrecked passengers off the coast of Newfoundland by using Marconi's technology. As Britain's postmaster general said at the time, "Those who have been saved, have been saved through one man, Mr. Marconi … and his marvellous invention."

About this same time, Reginald Aubrey Fessenden, a Canadian who some believe was one of Canada's greatest inventors and possibly even the country's most creative person, was involved in similar experiments that were designed to transmit human voices over long distances by radio waves. Fessenden was born in East Bolton, Quebec in 1866, and went to New York City in 1886. He eventually ended up working with Thomas Edison at his laboratory in West Orange, New

Jersey. While Fessenden never received the credit he deserved for his inventions, many communications experts believe he played the most important role of all in the development of the radio. In fact, many believe that radio broadcasting actually began with Fessenden on December 23, 1900, despite some earlier initiatives and developments in this area. Fessenden was working in his research facility on Cobb Island in the Potomac River in the United States that day when he hooked up a microphone to Morse code equipment in his laboratory and observed that it was snowing. What made this comment so historic was the fact that his assistant heard Fessenden's words and voice in Arlington, Virginia many miles away and confirmed by code that it was indeed snowing. Thus many experts in the communications field consider Fessenden to be the "father of radio broadcasting."

This claim was reinforced when Fessenden was the first person in the world to broadcast a program over the radio on Christmas Eve, 1906. Fessenden beamed out a Christmas concert using a rotary-spark transmitter and a 400-foot tower linked to his research facility at Brant Rock, Massachusetts on the Atlantic coast at precisely 9 o'clock. The concert included a short speech by Fessenden at the beginning, the playing of Handel's *Largo* on an Edison phonograph, Fessenden singing "O Holy Night" and playing it on the violin, and reading a passage from the Bible. Ships from all over the north and south Atlantic responded to this accomplishment by expressing their amazement over the first wireless broadcast of a radio program anywhere in the world. It was the first transmission of what is now known now as amplitude modulation or AM radio. The world has never looked back or been the same.

What made radio so important was the fact that it was

a *collective communications medium* that brought people together in millions of living rooms and kitchens to enjoy radio programs. Whereas the telephone was and still is more of a personal communications device (except for the once-popular party lines that allowed people to share lines and listen in on other people's calls and conversations), radio was far more of a collective medium from the very beginning. It possessed a remarkable capacity to bring people, families, and countries together in space and time as well as in private, public, and political terms. The possibility of sharing news, information, ideas, and various forms of entertainment proved impossible for people, marketers, corporations, governments, and politicians to ignore or resist.

Things moved very quickly after that. It didn't take long for radio to take off in terms of mass production and sales in Canada, the United States, Europe, and other parts of the world. It received another boost when Edward Samuel (Ted) Rogers—another inventive Canadian working in the communications field—created the first commercial alternating-current radio tube in 1925, as well as radios that plugged into a wall socket rather than depended on batteries that tended to wear out.

The contributions of Fessenden, Rogers, and others were among Canada's many creative contributions to the development of radio in its early years. Those contributions continued with the establishment of the Canadian Broadcasting Corporation (CBC) in 1936. While the CBC linked Canadians together from coast to coast to coast, it also had a significant impact on other parts of the world by linking Canada to the world and the world to Canada.

It wasn't long before Canada was known throughout the world for its International Service and National Farm Radio

Forum broadcasts and programs. The CBC's farm broadcasts began on the English and French networks in 1938 and 1939 respectively, and the National Farm Radio Forum began in 1941. Its International Service began in 1942 and was renamed Radio Canada International in 1970.

The farm broadcasts and the National Farm Radio Forum were designed to address domestic needs and problems in agriculture and farming. They provided educational programs and information to farmers in conjunction with the Canadian Association for Adult Education and the Canadian Federation of Agriculture. However, in 1952, the National Farm Radio Forum and its format were studied carefully by the United Nations Educational, Scientific, and Cultural Organization (UNESCO). This led to its format, structure, techniques, and educational programs being adopted by UNESCO to help farmers and agricultural officials in India, Ghana, France, and other parts in the world. This move proved to be very beneficial and popular in many parts of the world—and not just in Canada—and was a great hit with its motto "Read, Listen, Discuss, Act." These services are still provided today to farmers in Africa and other parts of the world, but on a voluntary basis through the Canadian charity Farm Radio International, which was created by a former CBC radio farm broadcaster named George Atkins in 1979.

The CBC International Service was designed to provide news and information to Canadian troops towards the end of the Second World War. In 1946 it also began to broadcast in Czech, Dutch, and Swedish to Czechoslovakia, Holland, and Sweden as well as to Latin America through its Sunday night broadcasts to Cuba, Colombia, Peru, Ecuador, and Brazil; Spain and Portugal in Spanish and Portuguese in 1947; and to the United Nations in 1952. By this time, the Service was fully engaged in the Cold War, with its signals and programs often jammed in the USSR and Eastern Europe. After its name was changed to Radio Canada International in 1970, it began broadcasting in Mandarin to more than 20 million listeners in Beijing, Shanghai, and Guangzhou in China who were anxious to learn English and thrived on its educational programs in this area. This was followed in 1971 with broadcasts in Arabic to the Middle East. The CBC and Radio Canada International were very popular in these and other parts of the world due to the high quality of their programming and the fact that they provided a more objective and impartial point of view than many American and European shortwave and radio stations that listeners found to be too biased.

In recent decades, radio has given way to television and other more modern forms of communications in influence and prominence, both in Canada and elsewhere in the world. Nonetheless, CBC Radio remains very popular with many Canadians, and in many cities (including Toronto) the CBC's "morning drive" radio program tops the ratings.

While newspapers have a much older history than radio or the telephone as communications media, their development has also been driven by the need to transmit information and ideas, this time in the form of the printed word and pictures. Here, as well, Canadians were in the forefront of some early, very important, and timely international developments.

Two Canadians—Georges-Edouard Desbarats and William Leggo—made a number of creative contributions to the international development of newspapers in the late nineteenth century. Having backgrounds in printing, they created the *Canadian Illustrated News* in 1869. It was the first newspaper in the world to use halftone photographs rather than engravings through a technique that Desbarats and Leggo invented. Capitalizing on this, they went to New York in 1873 where they founded the *New York Daily Graphic*. It was the first daily newspaper in the world to use photographic images, thereby adding a great deal of enthusiasm, excitement, realism, and intimacy to news coverage through the new technique of photojournalism.

Another Canadian who made a major contribution to the development of newspapers with international consequences at a crucial time was Frederick George Creed. He was born in Mill Village, Nova Scotia in 1871, taught himself cable and landline telegraphy, and decided to gain experience in this field in other parts of the world, most notably by working for the Central and South American Telegraphy and Cable Company in Peru and Chile, as well as in Scotland and England. He created a system for dramatically speeding up the transmission of Morse code through the invention of the Morse keyboard perforator. He also developed two other devices that had a major impact on newspaper production throughout the world—a transmitter-translator and a printer—that led to the creation of the Creed High Speed Automatic Printing Telegraph System. In 1920, the Press Association set up a private network that used several hundred Creed teleprinters to serve practically every daily newspaper in the United Kingdom. It wasn't long before companies in Australia, Denmark, India, South Africa, and Sweden followed suit. By the time the system was fully operational, it was printing information in a fraction of the time required by other systems and transmitting entire newspapers from London, England to major centres in Europe and many other parts of the world.

These developments were enhanced even more in the 1920s

when William Stephenson from Winnipeg, Manitoba created the wireless photo transmitter, making it possible to transmit pictures by radio waves or over telephone lines to any location in the world. This technology was instantly adopted by newspapers in most parts of the world, including the *Daily Mail* in London, England, which published the first international news photo in the world this way. This was not the only creative contribution Stephenson made to communications and the transmission and sharing of knowledge, information, and ideas. He also played a pioneering and strategic role in coordinating British, American, and Canadian military intelligence during the First World War.

With important precedents like these, it is understandable why Canadians played a powerful role in the development of some of the largest and most popular newspapers in the world, especially in Great Britain in the middle part of the twentieth century. The Canadians who performed this role were and still are well known in the newspaper business: William Maxwell Aitkin or Lord Beaverbrook, who was discussed briefly in Chapter 1; Roy Herbert Thomson, the First Baron of Fleet; and Kenneth Thomson, the Second Baron of Fleet. They were all in one form or another very controversial and provocative individuals throughout their lives and careers.

As important as newspapers have been and remain, both in Canada and other parts of the world, developments were afoot in the middle and latter part of the twentieth century to create other communications devices, vehicles, and systems capable of transmitting information, ideas, and messages at far faster rates as well as in a more efficient and effective manner. Once again, Canadians were in the forefront of some of the most important international developments in this domain, especially with respect to the creation, evolution, and use of satellites and wireless technologies that have had a substantial impact on the world over the last half century. For example, Canada's first satellite, Alouette 1, was launched by NASA in 1962 to orbit the earth and study the ionosphere. Following this, the Government of Canada created Telesat Canada in 1969 to operate the country's domestic commercial satellite system. This led to Canada being the first in the world to employ satellites for long-distance communication within a single country, a task which had previously been accomplished through large and expensive microwave relay networks. It was the need to service the large and sparsely populated Canadian North that made Canada an innovator in this field.

What is true for satellites and long-distance communication is also true for wireless communications. One of the most popular and important contemporary wireless devices in the world until very recently—the BlackBerry—was invented by Canadians. This occurred in 1999, fifteen years after Mike Lazaridis and Doug Fregin created Research in Motion (RIM) in Waterloo, Ontario in 1984. RIM grew rapidly after a great deal of experimentation, especially after Jim Balsillie joined the company as a dynamic administrator and marketing expert. With sales of its popular handheld wireless devices running into the billions of dollars, RIM was described by one author as being "as Canadian as maple syrup, the moose and hockey." Despite this initial success, the organization fell on hard times because of stiff competition from American communications giants such as Apple and Google, although more recently it is making a comeback through the development of communications software, new technologies for smart cities and automobiles, and other areas.

Interestingly, it was also a Canadian—James Gosling—who

invented the Java programming language, which is now the standard language used in many computer systems throughout the world. This occurred in 1994 when Gosling was working at Sun Microsystems in the United States. Add this to the fact that Canada has more than 30 million mobile users who send over 100 billion text messages a year and it is easy to appreciate the claim made by Bruce Powe, a well-known Canadian communications expert, that Canada was the first "wired nation" in the world and Canada is a "communications culture" first and foremost.

One would expect a country like Canada that has been challenged by many complex geographical, temporal, and spatial problems over the course of its history to have also produced some of the world's most creative and influential communications theorists and experts. The country doesn't disappoint in this area as well. Three of the most important are Harold Innis, Marshall McLuhan, and Dan Tapscott.

Harold Innis was born in Otterville, Ontario in 1894 and attended McMaster University and the University of Chicago. He is known throughout Canada for his innovative theories on the nature and functioning of the Canadian economy, as well as the powerful role communications plays in historical and contemporary developments throughout the world.

Innis's research and writing on the Canadian economy led him to develop a "staples thesis" based on the premise that Canada's economic and domestic development over the centuries has depended largely on the exploration, exploitation, and export of staples, commencing with fish and fur and continuing with lumber, wheat, gas, oil, potash, and many minerals. This theory continues to resonate as many economists today believe that Canada—and other nations in the world that are heavily dependent on staples—run the risk of slipping into what Innis called "the staples trap," namely too much dependence on staples and too little industrial and manufacturing development.

Later in life, Innis turned his attention to developing a number of far-reaching communications theories that are very relevant to past, present, and future developments in Canada and generally throughout the world. His writings on this subject focused on the role, functions, and impact that different types, modes, and systems of communications have on the development of people, countries, cultures, civilizations, and the world as a whole.

In specific terms, Innis contended that different media have different qualities, characteristics, and effects depending on whether they have a "time bias" or a "space bias." Time-biased media such as stone and clay tablets, handwritten manuscripts, and oral sources of information, for instance, last for a very long time but have limited reach in terms of the size of their audiences. Space-biased media such as radio, television, and the mass production and distribution of books and newspapers have much shorter "exposure times" but reach many more people over longer distances and larger areas. Innis concluded from this that time-biased media tend to favour stability, community, tradition, religion, decentralization, and hierarchical institutions whereas space-biased media tend to favour change, materialism, secularism, empire, centralization, and systems of government that are less hierarchical in nature.

Innis felt that a fine balance must be achieved and maintained in all countries of the world between the different media and the two main forms of communications. He warned that western civilization is in danger of losing this balance due to powerful advertising-driven media that are obsessed with "present-mindedness" and consequently "the continuous, systematic, ruthless destruction of elements of permanence that are imperative for cultural survival and progress." These and other theories are set out in two of Innis's most important books on this subject, namely *Empire and Communications* and *The Bias of Communications*. As such, Innis was a true pioneer in this rapidly-evolving field.

The intellectual tradition commenced by Harold Innis was expanded, enhanced, and intensified by his wife, Mary Quayle Innis, who also had a Ph.D. and was a prolific author and editor of some 14 books in addition to many short stories and magazine articles. As Harold Innis's fame grew, what became apparent was that his wife Mary had been actively involved in the development and publication of his books and theories on the impact of staples and communication media on Canadian and international development as well. In addition to this, Mary revised and edited *The Fur Trade*, *The Cod Fisheries*, and *Empire and Communication*, directly influencing the way Harold Innis and his theories are read and interpreted today, as well as collaborating with her husband on the research and writing of *An Economic History of Canada*. Mary also wrote a two-volume book called *Changing Canada* published in 1951-52, *The Clear Spirit*, a biographical account of 20 prominent women published in 1966, several other books, and, in addition, served as Dean of Women at University College of the University of Toronto from 1955 to 1964.

Interestingly, the curiosity, capabilities, and creative achievements demonstrated by Harold and Mary Quayle Innis have been sustained by their children and especially Anne Innis Dagg well into the twenty-first century. Anne was among the first women in the world to study animal behavior in Africa in the wild, along with Jane Goodall and Dian Fossey. Her particular interest was giraffes, which she became fascinated with after seeing them in a zoo in her childhood. In the 1950s, she realized a long-held dream when she went to Africa to study giraffes in depth in their natural habitat, which later led to the publication of the classic book *The Giraffe: Its Biology, Behavior, and Ecology* (co-authored with J. Bristol Foster, 1976). Both a zoologist and active feminist, Dagg has spent her life studying the behaviour of animals in zoos and in the wild, as well as fighting for women's rights in academic institutions across Canada and elsewhere in the world. The 2018 documentary *The Woman Who Loves Giraffes* tells the story of Anne's "giraffe journey" and of the hurdles she faced in her career, and includes film footage and photos from her first trip to Africa as well as a more recent visit when she and her daughter returned to many of the same places. While it has taken many years, Anne is finally being recognized for her outstanding accomplishments in zoology, biology, and feminism, thereby expanding on the remarkable intellectual tradition commenced by her parents.

But back to Harold Innis for a moment longer. His innovative theories in communications did a great deal to pave the way for the emergence of another outstanding and highly creative Canadian scholar, Marshall McLuhan, without doubt one of Canada's and the world's greatest communications theorists and experts. McLuhan is generally regarded as someone who foresaw the development, dominance, and power of contemporary communications, long before the social media revolution now sweeping the world. What Steve Jobs was to the development of modern communications technology and the iPhone, Marshall McLuhan was to communications theory and the development of the electronic era.

Like Innis, McLuhan was fascinated with the different

qualities and characteristics that the media possess, as well as the impact they have on people's lives as well as on social, political, and cultural developments and change throughout the world. Three of his most important books on these matters—*The Gutenberg Galaxy: The Making of Typographic Man* (1962), *Understanding Media* (1964), and *The Medium is the Massage: An Inventory of Effects* (1967)—were best-sellers and had a significant impact on people and countries in all parts of the world.

Believing that the print era was in decline and the electronic era was on the rise, McLuhan famously claimed that "the medium is the message," not the contents of the medium. The nature of specific communications media have a powerful effect on people's sensory and brain development according to McLuhan. As Bruce Powe, mentioned earlier and a student of McLuhan's, observed, "The recent research on the so-called iBrain … is all anticipated in McLuhan."

But McLuhan's fascination with the impact of the electronic media went much farther and deeper than this. Whereas the print media produce a "linear concept of time" and a "one thing after another" approach to development, the electronic media produce a "simultaneous concept of time" and an "all-things-at-once" approach to development. Thoughts like this led McLuhan to conclude that the world was rapidly becoming a "global village" with a diversity of "hot" and "cold" media. This led him to conclude that education should be radically transformed. It should no longer be concerned with formal lectures and the teaching of "definitive truths," but rather with lively discussions and debates between teachers and students, joint explorations and discoveries of ideas, and becoming literate in many different media rather than only the print media.

As McLuhan's ideas began to take hold inside and outside the educational system, his reputation grew exponentially in North America, Europe, and other parts of the world. In the 1990's, *Wired* magazine called McLuhan the "patron saint" of the Internet. This was confirmed by Robert Logan, a physics professor emeritus at the University of Toronto and a close friend of McLuhan's, who said that McLuhan's theories and writing anticipated numerous developments in the Internet and communications, from Wikipedia and Twitter to laptop computers and smartphones. This was reinforced even more when *The Guardian* and *New York Times* included a book by McLuhan on their list of the 100 greatest non-fiction works of all time.

Talk about having a major impact on the world! In a prophetic interview with *Playboy* about this time, McLuhan said, "In the electronic age of instantaneous communications ... our survival, and at the very least our comfort and happiness, is predicated on understanding the nature of our new environment. If we understand the revolutionary transformations caused by new media, we can anticipate and control them; but if we continue in our self-induced ... trance, we will be their slaves."

What McLuhan was to the electronic revolution that swept the world in the latter part of the twentieth century and early part of the twenty-first century, Don Tapscott is to the digital revolution going on in the world today. In two of his earlier books—*Growing Up Digital: The Rise of the Net Generation* and *Grown Up Digital: How the Net Generation Is Changing Your World*—Tapscott made the case that young people who were born between 1977 and 1997—"Net Geners" as he calls them—are really the first generation to grow up completely in a digital world that is totally different from the world of their baby-boomer parents and the McLuhan generation. Whereas the latter generation tended to watch a great deal of television, Tapscott contends that the Net generation is busy spending its time buying and using cellphones, smartphones, and interacting online, and consequently on scrutinizing, authenticating, and organizing information, sharing thoughts and ideas, texting, and collaborating in one form or another. This, in Tapscott's opinion, is leading to the creation of new types of people as well as new institutions, technologies, systems, and structures.

This concludes our examination of the way Canadians have dealt with their vast and varied transportation and communications problems, producing creative and productive solutions to them over the last few centuries and especially over the last 50 years. It also concludes our examination of the powerful effect many of their initiatives and inventions have had on people and countries in other parts of the world and on the world as a whole, not just on Canada and Canadians. These initiatives and inventions have resulted from the necessity of attending to one of the most fundamental human requirements of all, namely the need people, societies, and countries have to connect, communicate, interact, and survive. Without the innovative achievements discussed in this chapter, it is very likely that Canada would not exist as a country and the world would be much different than it is today.

Chapter Three
Resources, Agriculture, and Industry

It is a well-known fact that Canada's economic development over the centuries has been focused largely on natural resources and basic staples. In earlier times, fish, fur, timber, wheat, and kerosene (derived from petroleum) predominated. More recently, oil, natural gas, coal, minerals, and hydroelectricity have risen to prominence, although such staples as fish, wheat, and timber remain important.

Each of these natural resources and basic staples has given rise to a specific stage in Canadian development focused on the exploration, extraction, processing, utilization, and export of the resources and staples in question. As well, each stage has required a great deal of creativity to make these resources and staples available for domestic consumption and international markets.

Whereas the Indigenous people used the resources of the country primarily for survival, consumption, and local trading purposes, the European and other immigrants who came to what eventually became Canada used the country's resources not only for these purposes but also for export to other parts of the world. In fact, their survival and the survival of the communities and colonies they created would not have possible without such export, as well as the assistance that was provided by the Indigenous peoples in helping them to come to grips with the demands and dictates of living in an unknown land with many exceedingly difficult problems and obstacles to overcome.

Fish was the first resource to be exploited by the European

and other settlers for international as well as domestic use. This was especially true for fish that could be stored in dried or salted form for many months and transported over long distances without too much difficulty. As early as the sixteenth century, Spanish, Portuguese, French, and English fishermen were busy fishing in the Grand Banks off the coast of Newfoundland for cod and other fish that existed in vast quantities there. They were able to capitalize on the strong demand for fish in Europe, especially southern Europe, which was predominantly Catholic and the demand for different types of fish was substantial.

This provided an excellent opportunity to develop the fishing industry in Canada and especially the exporting of fish to Europe and other parts of the world. In the centuries that followed, a large fishing industry was created and flourished on the east and later west coasts of the country. With this came the establishment of numerous fishing villages, the construction of schooners, fishing fleets, and processing plants, and the harvesting of vast quantities and varieties of fish. Canada continues to have a lucrative fishing industry today and exports huge quantities of ocean, freshwater, and shell fish of a very high quality to many parts of the world: Atlantic and Pacific salmon, crab, Arctic char, clams, oysters, mussels, eel, lobsters, cod, halibut, mackerel, sturgeon, gold eye, white fish, pickerel, bass, pike, trout, and others. Many of these delicacies are packed up and shipped off in fresh, smoked, or frozen form to destinations around the world. The importance of being able to package fish for export helps to explain why it was Canada and a Canadian—Dr. Archibald Huntsman—who first developed frozen food for commercial use when he created frozen fish fillets while working as a marine scientist in Halifax at the Biological Board of Canada (now the Fisheries Research Board).

Strange as it may sound, the fishing industry gave rise to the fur trade, the next great resource to be tapped in substantial quantities in Canada for international and domestic use. The establishment of permanent fishing villages in the Maritimes spurred the trade in furs because the fishing villages needed an additional source of revenue if they were to survive in the cold climate and punishing geography. The European and other settlers discovered that the Indigenous people were very skilled at catching and skinning animals, and were very anxious to trade the fur that resulted from this for knives, kettles, axes, blankets, and other commodities. Once again, Europe provided to be an ideal market for the country's exports. When the demand for wide-brimmed fur hats and winter coats skyrocketed in Europe in the seventeenth century, fur traders, merchants, and the Indigenous peoples moved quickly to fill this demand and capitalize on this opportunity. This was particularly true for the export of beaver pelts, which made the finest hats and coats because the fur was soft and had a wonderful matted quality, especially if it was acquired in the winter when the fur was at its thickest and finest.

While the state of the fishing industry has been relatively stable for the most part over the centuries (except when fish stocks in the Grand Banks and elsewhere were overfished and depleted), the history of the fur industry has been much more erratic and controversial, especially in recent years. This results from resistance in Canada and other countries to killing animals for their fur, especially the killing of young seals. Strong opposition to this practice has escalated in Canada and elsewhere as attention was drawn to the killing of countless seals and the inhumane way in which they are often killed.

Just as there was a huge demand for fur from Canada in most

parts of Europe in the sixteenth, seventeenth, and eighteenth centuries, so there was a huge demand for timber in Europe and other parts of the world in the eighteenth and nineteenth centuries.

What fuelled this demand was the need for masts for the British Navy, the largest and most powerful navy in the world at that time. Not only was Britain locked in a fierce battle with Spain, Portugal, France, and Holland for supremacy on the high seas, but it was also heavily engaged in North America and the West Indies, actively involved in the American Revolutionary War, the leader of the coalition opposing France during the Napoleonic Wars, as well as passing through a period of profound economic transformation and social change.

While Britain had a good supply of certain types of timber, it was deficient in pine, which made the best masts for ships and was available in large quantities along the Atlantic seaboard. After relying on timber from the Baltic for many years for its masts and other requirements, Britain turned to North America to supply it with the timber it needed after its supply from the Baltic countries proved insufficient, erratic, and complicated by all sorts of political and transportation problems. This caused a huge surge in the demand for timber for masts and other wood products, especially shingles, barrel staves, boxes, and spool wood for the textile industry. By 1810, the timber trade accounted for more than 70 percent of the exports from British North America.

By this time, Nova Scotia and New Brunswick were largely depleted of their timber resources and regions in Ontario and Quebec far from river valleys were being exploited for this purpose. Moreover, British Columbia was rapidly becoming one of the largest suppliers of timber in the world. This occurred primarily after 1850, when prospectors turned their attention to the province's vast timber reserves after the Fraser River gold rush had run its course. Not only were there millions of huge trees in British Columbia—Douglas fir, red cedar, Sitka spruce, hemlock, balsam, and so forth—but also the construction of the Canadian Pacific Railway in the 1880s reduced the cost of transporting British Columbian timber to markets in eastern Canada, Europe, and other destinations in the world. This made British Columbia one of the largest, most prosperous, and profitable timber-producing areas in the world.

Many creative developments were required to make timber and pulp and paper a success in Canada and other parts of the world. Prior to Confederation, many small paper mills were operating in Upper and Lower Canada, as well as in the Maritimes. They were engaged in making various types of paper, which at that time was made from rags. When the supply of rags became insufficient due to the growing demand for paper, the search was on for ways to make paper from other materials and substances.

As often happens, necessity proved to be the mother of invention in this case as well. In the 1830s, Charles Fenerty of Sackville, Nova Scotia, began experimenting with alternative ways of making paper. Since the vegetable fibre in paper was derived from the cotton and linen fibres found in rags, Fenerty believed that paper could be made from wood because it also contained the necessary fibre. Aware that the finest wood substance was the fuzzy, lint-like residue that was produced from the wooden saw frame rubbing against the wooden sides, he carefully collected and worked this fine wood residue by hand, using all the standard techniques of wetting, bleaching, molding, flattening, and shaping that were required to make paper. This enabled him to produce a small quantity of *pulpwood paper* on which the entire pulp and paper industry throughout the world depends today.

This by no means completes the story of Canada's creative contributions to this industry. In 1864, John Thomson of Napanee, Ontario created the first chemical wood pulp process. Then, in 1866, Alex Buntin created the first commercial mechanical pulp mill in North America at Valleyfield, Quebec. By the time of Confederation a year later, Canada had one of the largest, most creative, and lucrative paper-making industries in the world. In fact, Canada became the largest international producer of pulp and paper by the beginning of the First World War. And by 1929, Canada's pulp and paper companies were producing more than three million tons of paper a year, an incredible 65 percent of the world's total. This made the pulp and paper industry one of the most important industries in Canada, especially after the demand for newsprint escalated rapidly as a result of the mass popularity of newspapers in Canada, the United States, Europe, and other parts of the world.

Like the pulp and paper industry, Canada's wheat farmers also came to command a large share of world markets, though this happened in a totally different way and resulted largely from the invention of one of Canada's and the world's greatest agricultural products. This is Marquis wheat. It was created by Charles E. Saunders in 1908 and played a major role in "filling up the West," which Sir John A. Macdonald, Canada's first Prime Minister, felt was the solution to resisting imperialist pressures from the United States, as attracting immigrants from abroad to settle the Prairies would help to ensure Canada's survival as a nation. Doing so was not possible, however, without creating agricultural products that could withstand Canada's long winters, cold climate, early frosts, and short growing season, especially in the western provinces.

Saunders was a cerealist at the Dominion Department of Agriculture in Ottawa when he invented Marquis wheat. Prior to this time, the most common variety of wheat grown in Canada was Red Fife, a rust resistant strain created by David Fife in Ontario in 1843. While this particular brand of wheat served many parts of the country well where the growing season was longer, it was not viable in the west where the growing season was shorter—less than a hundred days in duration due to the early frosts. Capitalizing on research conducted by his father, Dr. William Saunders, who was also an extremely inventive person and first Director of the Central Experimental Farm in Ottawa (which in itself was a very creative achievement established in 1889), Charles Saunders developed Marquis wheat by crossing Red Fife with an Indian variety called Hard Red Calcutta. Since Marquis wheat matured in a shorter period of time, produced significantly higher yields, and was more resistant to disease, it proved to be a bonanza for western farmers, the Canadian economy, and eventually the entire world.

Not only did Marquis wheat mature a week to ten days sooner than Red Fife, but it also yielded five more bushels per acre. In 1911, it was awarded the gold prize at the World's Fair in New York as the best wheat grown in the world. As a result, Canada quickly became known throughout the world as one of the most important producers and exporters of wheat in the world by people who depended on this "miracle grain" from Manitoba, Saskatchewan, and Alberta. By 1914, Canada had expanded its Marquis wheat exports to many states in the American midwest, including North Dakota, Minnesota, Montana, Iowa, Nebraska, and South Dakota, due largely to its faster maturity and superb meal, taste, and milling and baking qualities. Three years later, Marquis wheat constituted 80 percent of Canada's total wheat acreage, covering more than 20 million acres and valued at more than half a billion dollars in revenue.

It was also being exported to many other parts of the world by this time, which helped Canadian allies during the First World War, especially the United Kingdom, France, Belgium, and Greece. By 1920, more than 90 percent of the 17 million acres of wheat produced in western Canada was Marquis, as was 70 percent of the wheat grown in the United States. This fact explains why many people at that time considered Canada to be the "bread basket of the world." Canada's exports of wheat and grains in general—and Marquis wheat in particular—to Europe, Russia, the Middle East, North Africa, and other destinations in the world have continued to grow substantially. Not surprisingly, Charles Saunders was made a fellow of the Royal Society of Canada in 1921 and knighted by King George V for his valuable contributions to world agriculture in 1933, thereby becoming Sir Charles Saunders.

While many people know about Marquis wheat and the creative ingenuity that was required to create and produce it, far fewer know about canola and the comparable amount of creativity its creation and production has necessitated, not only by farmers, scientists, universities, government agencies, commercial crushers, and wheat pools, but especially by B. R. Stefansson, Keith Downey, Burton Craig, and others who played a seminal role in its creation and evolution. Like Marquis wheat, it is a miracle foodstuff that was originally produced on the Canadian Prairies and especially Saskatchewan.

Rapeseed, from which canola is derived, had been known for centuries. However, it was a low-acreage crop because it was deemed to be toxic in character and have poor meal qualities. Even after it was discovered that rapeseed could produce a useful industrial oil that clung well to metals, production was limited. In Canada, for example, acreage had diminished from 20,000 acres to 400 acres by 1950.

It was about this time that Stefansson, Downey, Craig, and others arrived on the scene. Believing that rapeseed possessed enormous potential, they experimented for many years with various types of rapeseed—much as Charles Saunders did with various types of wheat—until they produced a LEAR variety in 1964 that began the transformation of rapeseed into canola. By 1977, they had solved the two basic problems encountered with rapeseed—high erucic acid and too much glucinaolate—and created an excellent variety of canola that yielded a high-quality oil that had excellent meal properties. It is grown largely in countries that have dry climates and short growing seasons, although its commercial use has been escalating rapidly in all parts of the world in recent years due to the fact that it is low in saturated fat, high in monounsaturated fat, and smokes very little when heated in a frying pan. This new variant of rapeseed was named "canola" because the word sounded like "Canadian oil" to its creators.

By 1981, these inventive individuals had turned rapeseed into pure gold, dramatically changing the colour of the Canadian prairies in June to a brilliant yellow when the canola crop ripens. They also produced a multi-billion dollar industry that proved that Canadians are capable of taking commercial advantage of some of their most creative achievements after several devastating experiences in this area in the nineteenth century. Canola is now grown on roughly 8 million hectares of farmland in Canada. In 2016, Canadian farmers harvested canola worth about $8.6 billion, making it the most valuable field crop by revenue produced in the country. Production had doubled in ten years and increased to more than 18 million tonnes by 2016 due to the high demand for canola oil in other parts of the world. As a result, canola is now the third-largest source of vegetable oil in the world, after soybean and palm oil, and is used extensively throughout the world as a major crop for edible-oil production, although some researchers still contend that it possesses some toxic residues because of its derivation from rapeseed. Nevertheless, as stated in a recent National Research Council publication, *From Rapeseed to Canola: The Billion Dollar Success Story*, "Wheat may always be king, but canola wears a crown of considerable weight and undisputed beauty."

At present, major producers of canola products—in addition to Canada, which accounts for about one-half of the world's trade in canola seed, meal, and oil—are China, India, and the U.S. Major importers are Japan, Mexico, China, the European Union, and the U.S. In 2019, however, China barred imports of Canadian canola seed into that country due to claims of pest infestations found in Canadian seeds, although Canadian authorities believe this action was in response to the arrest of senior Huawei executive Meng Wanzhou in Vancouver in 2018 on an extradition warrant issued by the United States. This move was very troubling to producers in Canada since in 2018 sales to China accounted for 40 percent of Canada's canola production, valued at approximately $ 2.7 billion. Even more troubling was China's subsequent banning of imports of Canadian soybeans, peas, pork, and beef, which resulted in substantial losses for Canada's farmers and agricultural industry.

It is doubtful if Canada's experiences with canola products

and Marquis wheat would have been as successful and profitable as they have been if it had not been for the invention of the self-propelled combine harvester. While combine harvesters had been in existence for a long time and were very effective because they combined three operations in one—reaping, threshing, and winnowing—the *self-propelled* combine harvester invented in Canada represented a considerable improvement on this. It was created by Thomas Carroll in 1937 when he worked at the Massey-Harris Company, and was improved even more when a lighter-weight version became available and was marketed by Massey-Harris in 1940. Since this new device dramatically increased and quickened the harvesting of many grains, it wasn't long before Europeans picked up on this Canadian invention. In 1952, Claeys launched the first self-propelled combine harvester in Europe, and in 1953, the European manufacturer CLAAS developed a self-propelled combine harvester called "Hercules" that could harvest up to five tons of wheat a day. Nevertheless, this does not alter the fact that it was a Canadian invention that triggered these subsequent developments and is now used wherever the growing of grains and production of cereals in the world is prevalent.

Marquis wheat and canola are not the only Canadian agricultural products that have had a profound effect on the world. Believe it or not, another is the potato, but in manufactured form rather than its natural form. While few people are aware of it, Canada is the largest producer of french fries in the world as a result of this.

It all started in 1823 when several members of the McCain family emigrated from Scotland to Ireland and then came to New Brunswick in eastern Canada in search of a better life. More than a century later, two descendants of the original immigrants, third-generation farmers Harrison and Wallace McCain, founded McCain Frozen Foods in 1956 and built it into one of the largest frozen food companies in the world, basing its development on such key values as authenticity—"be real, open, and informal"—commitment—"create your future and be the vision"—and trust—"do the right thing."

Predicated on these values, McCain Frozen Foods remains a major player in the global frozen food and food processing industry today. The company has provided jobs for more than 17,000 people worldwide, generated global sales in the $10 billion range, overseen the construction of more than 41 sites on six continents, and processed millions of tons of potatoes. Although the company's base plant is located in a small town in New Brunswick with some 2,000 inhabitants—Florenceville-Bristol—it has been able to capitalize on one of the best potato-producing areas in the world, namely New Brunswick, Nova Scotia, and especially Prince Edward Island. As a result, people from all over the world now enjoy McCain's french fries, and possibly poutine as well. This is french fries smothered in cheese and gravy, a concoction created by Fernand Lachance in Quebec in 1957 that has since grown rapidly in popularity in many parts of the world.

While we are discussing agricultural products, what about two extremely refreshing Canadian beverages that are now very popular throughout the world and exported to many countries in the world on a regular basis? The first is Canada Dry Ginger Ale, and the second is ice wine.

Ginger ale is certainly one of Canada's and the world's best-known drinks. It was invented by John J. McLaughlin, a chemist and pharmacist who produced flavoured extracts for the soda water bottling plant he created in Toronto that began mass producing soft drinks for sale beyond the traditional soda fountains in drug stores. In 1904, this company introduced a Pale Dry Ginger Ale with a label depicting a beaver sitting on top of a map of Canada, which has since been replaced by a crown. In 1907, McLaughlin trademarked his famous Canada Dry Ginger Ale and started manufacturing it commercially at a plant on Sherbourne Street in Toronto, calling it "the champagne of ginger ales."

When McLaughlin began shipping this drink to the United States, it became an instant success, so much so that he opened a plant in Manhattan shortly after this. It became extremely popular as a mixer during Prohibition in the United States when it was discovered that its flavour helped to mask the taste of homemade (illegal) liquor. In the 1930s, Canada Dry expanded worldwide and now has major manufacturing plants in many countries and parts of the world, including the United States, Mexico, Colombia, Peru, the Middle East, Europe, and Japan in addition to Canada. While the name of the beverage remains Canada Dry, the brand has been owned by many different companies since 1923, and is produced and distributed by Keurig Dr Pepper today.

Turning now to another popular Canadian beverage, while Canada started to make wine commercially much later than most European countries, it has developed a very successful wine industry over the last few decades with major wineries in British Columbia, Ontario, Quebec, and other parts of the country. These wineries are winning many awards and competing effectively with European, Australian, Argentinian, American, and other major wine producers. Canadians have become well-known throughout the world for their creativity in making wine, not only as a result of their ability to blend domestic and foreign grapes to produce some the best wines available, but also for the innovative contributions they have made to the production of ice wine. In making this wine, Canadians have manifested a remarkable capacity for taking advantage of specific climatic, geographic, and topographical conditions and turning them to their advantage. Ice wine is produced by picking the grapes at a very specific time of year, usually after a hard freeze, and then crushing and aging them effectively. The specific climatic conditions necessary for producing this wine have made Canada one of the world's top producers of ice wine, along with Germany.

What holds true for the creative contributions Canada has made to the world through resources, staples, and agricultural products such as these and many others also holds true for industry and industrial products. One of the first individuals to achieve success in this area was Abraham Gesner. He was born in Cornwallis, Nova Scotia in 1797 and was keenly interested in geology and mineralogy from a very early age. He studied and mapped the geology of Nova Scotia between 1827 and 1837, and set out his findings in a remarkable book entitled *Remarks on the Geology and Mineralogy of Nova Scotia*. He also worked as the official geologist of New Brunswick for several years before turning his attention to the creation of a better source of indoor illumination.

Up to that time, artificial light was produced by burning different substances, most often animal and vegetable oils, wood, and wax. The problem was that light from these sources was feeble, smoky, smelly, and dangerous. What Gesner did through his numerous experiments in this area was produce a far superior source of light. He did this through the development of kerosene in 1846, as well as inventing the process whereby kerosene could be extracted from coal and petroleum and manufactured commercially to produce a beautiful bright yellow light. Not only did this revolutionize life in Canada but it did so elsewhere in the world. Gesner obtained patents for the production of kerosene in 1854 and oversaw the creation of a factory on Long Island, New York to manufacture kerosene commercially. Kerosene soon became the standard lighting source for many homes in the United States and other parts of the world and not just Canada after this.

Gesner influence was not limited to kerosene. Far from it. Not only did the discovery of this lighting source revolutionize indoor lighting, but also it provided the impetus that was needed to develop the petrochemical industry, one of the largest and most lucrative industries in the world today. Indeed, many consider Gesner to be "the father of the petrochemical industry," since the production of kerosene from coal and petroleum eventually led to kerosene being used not only for lighting, heating, cooking, chemical, and transportation purposes, but also for the production of plastics, fertilizers, detergents, paints, textiles, dyes, solvents, and many other products.

In addition to this, Gesner played a seminal role in the creation of the North American Kerosene Gas Light Company in New York, which was later acquired by Charles Pratt and renamed the New York Kerosene Company. Pratt eventually joined forces with John D. Rockefeller in the creation of Standard Oil, which quickly became the largest oil company in the world. The successors of this company—ExxonMobil and Chevron—are still among the world's largest companies, and, of course, Imperial Oil in Canada, which was formed as a subsidiary of Standard Oil, still markets its gasoline under the Esso (from "S.O.," the initials of Standard Oil) brand.

Many of the techniques Gesner developed for kerosene and petroleum production were spelled out in detail in his book—*A Practical Treatise on Coal, Petroleum, and other Distilled Oils*—which was deemed to be the Bible in the petrochemical and oil industries for many years.

About the same time that Gesner was making his discoveries and leaving his mark on the use of kerosene and petroleum in North America and elsewhere in the world, other developments were taking place in Canada and other parts of North America that would eventually vault oil into a prominent position in the world economy. In 1854, the same year that Gesner acquired his patent for kerosene, Charles N. Tripp of Woodstock, Ontario created the first oil company in North America. When this company was taken over by James Miller Williams, the first commercial oil well in North America and arguably the world was drilled at Oil Springs, Ontario in 1858. In doing so, Williams established Canada's reputation as a global leader in the oil and gas industries.

Williams was born in United States but came to Canada in his early twenties shortly after the War of 1812. He was originally trained as a carriage maker, and established the Hamilton Coach Factory with H. G. Cooper when he moved to Hamilton in 1846. The company was very successful, largely because it manufactured railroad cars for the new Great Western Railway that travelled from Toronto through Hamilton and London on its way to Detroit in the United States.

Believing that oil was the fuel of the future, Williams acquired land in Enniskillen, northwest of London, and eventually struck oil at Oil Springs in 1858. He went on to build a highly successful business refining and shipping oil through the Canadian Oil Company he founded. As confirmation of the pioneering role he played in the creation of the oil industry, he is identified in the U.S. geological records in Washington, D. C. as "the discoverer of petroleum and the pioneer in its refinement and preparation for illuminating and lubricating purposes." Just how important this was for Canada, the United States, and the entire world is revealed in the following observation by Murray White on the "Oil Industry's Birthplace":

> Oil Springs today is a unique perch from which to view the oil industry's impact on modern society—without it, there wouldn't be one.
>
> The discovery of oil here, and its first commercial application as lamp oil in 1858 by James Miller Williams, seeds a global revolution: A cheap, plentiful energy source had been found, and in the century that followed, it would enable the rapid growth of the industrial era.
>
> Oil allowed for the engineering of combustion engines, which would replace steam power as powerful, efficient drivers of transportation, from trains to ships to, eventually, cars and planes.
>
> With it came complete social transformation: geography disappeared as means of transportation got faster and faster. Global travel and commerce became quick, efficient, cheap. Urban centres dwindled as cars transported more and more people into suburbs. The automotive industry created empires. With it, oil wealth, and its pursuit, proliferated.[5]

Williams' remarkable achievements were merely the first in a long line of creative achievements by Canadians in the oil and

---

5. Murray White, "Oil industry's birthplace," *Toronto Star*, May 11, 2008, A1, A7.

gas industries. In 1862, the first of a number of oil "gushers" erupted in Petrolia that went on to achieve international status throughout the 1860s when the first oil pipeline was built in North America from Petrolia to Sarnia. It was claimed by locals in the area that the "jerker rod system" was invented at Petrolia. This made it possible to carry power to many widely dispersed oil wells by using a series of linked poles that were made to oscillate through a single power source. Interestingly, the expertise that was developed in drilling at Petrolia was utilized to great advantage in the development of oil fields as diverse and far away as Java, Galicia, Germany, and Hungary.

Canada never looked back as far as the development of the oil and gas industries were concerned. In the century following Williams' creative contributions, this included the discovery of oil in Alberta in the twentieth century, first at the Leduc field and then the Athabasca oil sands. These developments, along with many others, helped to propel Canada into the front ranks of the major industrial economies of the world, with levels of per capita income and industrial output right up there with the top nations of the world.

What Williams did for the oil and gas industry and Gesner did for kerosene and the petrochemical industry, Thomas "Carbide" Willson did for calcium carbide, acetylene, shipbuilding, the automotive industries, and the creation of an even brighter source of light than that given off by kerosene.

Willson was born in Woodstock, Ontario in 1861. While calcium carbide and acetylene were known at this time, Willson invented and patented a unique method whereby calcium carbide could be produced cheaply by heating coal, lime, and tar to extremely high temperatures, and acetylene could be produced equally cheaply by adding water to calcium carbide.

Willson's creative discoveries in these two areas are generally recognized as giving rise to the development of two giant commercial companies: the Union Carbide Company in the United States and the Shawinigan Chemical Company in Canada. These two companies dominated the field for many years, and were involved with Willson in the creation and development of the oxyacetylene torch between 1895 and 1904 that made it possible to weld heavy metals. This torch was capable of reaching temperatures of 6,000 degrees Fahrenheit, hot enough for welding and cutting metals and therefore facilitating key developments in the shipbuilding and automotive industries in Canada, the United States, and eventually many other parts of the world.

By this time, two Canadians—Henry Woodward, a medical student from Toronto and his friend, Matthew Evans—had been granted a patent from the Patent Branch of the Department of Agriculture in Ottawa—patent number 3738—for an electric light. In their patent application, they stated that they "jointly invented new and useful improvements in the art or process of obtaining Artificial light by means of Electricity." This occurred in 1874, making it possible to claim that Canadians played a seminal role in lighting up the world for a second time—the first through the discovery of kerosene by Gesner, and the second through the invention of electric light by Woodward and Evans.

Unfortunately, Woodward and Evans were unable to raise the funds that were required to produce electric light commercially in Canada. As a result, Woodward went to the United States where he applied for an American patent for this invention. He was granted a patent for electric light—American patent number 181613—on August 29, 1876. When he was unable to raise the funds needed to develop this invention in practical terms, he sold the rights to Thomas Edison, one of America's best known and most prolific inventors. As a result, Edison is generally credited throughout the world as the person who invented electric light. However, although Edison made numerous changes to the design of the electric light bulb that resulted in it being longer lasting and commercially viable, his work built on the earlier creative accomplishments of others, especially Woodward and Evans, something that Edison was never willing to admit.

For some curious reason, Canadians seem to have a real fascination with lighting and producing ever stronger, better, and brighter sources of light. As a result, Canadians were involved in lighting up the world not once, not twice, but a third time. This third time occurred when Thomas "Carbide" Willson paved the way for the creation of yet another type of light—a type of light that shone even more brightly than the light produced by incandescent electric light bulbs. It was light with a very bright and intense flame that was produced by burning acetylene. In addition to its first use as a lamp fuel for homes, acetylene was also used for headlights for cars, carriages, trains, and bicycles because it shone extremely brightly. It was also used for lamps on miners' helmets, lighthouses, and street lamps, indeed anywhere where high intensity illumination was required.

Most of the creative Canadians we have discussed thus far were able to see at least some of the fruits of their creativity while they were still alive. However, others were not so fortunate. Nevertheless, they created things that have also had a strong and lasting effect on the world. One of these individuals was Eric Leaver.

Leaver was born in England in 1915 and came to Saskatchewan at an early age. He established himself as a major inventor of automatic landing systems for aircraft in the late 1930s. After World War II, he founded Electronic Associates Ltd. in association with Research Enterprise Ltd. and began creating robotic machine tools as well as a functional "hand-arm machine." In 1946, he created a system known as the "Automatic Machine Control by Recorded Operation," or AMCRO. He wrote about this system and its implications for industrial production in an article published in the November 1946 issue of *Fortune* magazine titled "The Automatic Factory." In this article, Leaver set out his ideas for a fully automated factory, thereby becoming the first person in the world to visualize and foresee what a full-fledged automated system of industrial production would look like and how it would function, as well as to write about this in detail in a well-known and well-respected publication. While his AMCRO system was patented in 1947, unfortunately Leaver lacked the risk capital to develop it in Canada. As a result, it was developed in the United States by licensees, but was resisted by most manufacturers of conventional machine tools as being too threatening and therefore not developed fully at that time. However, it has received far wider applications since that time to say the least!

Undeterred by disappointments like this, Leaver went on to create a small plant in Toronto in conjunction with G. R. Mounce, a Canadian engineer, that produced the first production tool that could be programmed to repeat the operations of a worker making a product. This was the forerunner of automated systems that are now used frequently by industries in all parts of the world with no end in sight. It is a perfect illustration of the International Society of Automation's definition of automation as "the creation and application of technology to monitor and control the production and delivery of products and services."

Leaver and Mounce went on to produce many other electronic devices, including a portable Geiger counter for uranium prospectors, radar altimeters for aerial surveys, and automatic process controls for industry, most notably for pulp and paper mills and mines. However, these two inventive individuals are most widely recognized throughout the world as the founders of automation, one of the most important technological innovations in the modern world. Not only does automation have many applications in manufacturing, but also it plays an essential part in robotics and artificial intelligence, two rapidly expanding fields today.

In addition to the creative achievements already discussed, Canadians have also made seminal contributions to the development of many other industrial products and technological devices. While few Canadians or people in other parts of the world are aware of it, the electric oven, automatic lubricating cap, gramophone, caulking gun, Robertson screw and screwdriver, Plexiglas, paint roller, transmission electron microscope, green garbage bags, electronic wheelchairs, Ardox spiral nails, the alkaline long-lasting battery, walkie-talkies, biodegradable plastics, safety paint, gas masks, fathometer, airplane flight recorder ("black box"), and other devices and processes were either created by Canadians or Canadians played a seminal role in their development. While the impact of some of these products and devices has been confined primarily to Canada, others have had a major impact on the world as a whole. Three of the most important of these are the electric oven, the paint roller, and the transmission electron microscope.

Thomas Ahearn was undoubtedly one of Canada's most creative individuals. He was fascinated throughout his life with electricity and made many innovative contributions to its development and use. This included installing arc street lights in Ottawa, the production of street cars and creation of a street railway system in Ottawa, development of electric heaters to warm street cars, and the invention of a rotating brush to clear street car tracks during the winter months.

However, his crowning achievement was the invention of the electric oven. In 1892, Ahearn and his colleague Warren Y. Soper filed patent number 39916 for an oven of this type. As co-owners of Ottawa's Chaudière Electric Light and Power Company, they prepared the first meal ever cooked electrically on August 29 of that year at the Windsor Hotel in Ottawa. It was billed as an "Electric Dinner" because everything on the menu was cooked using electricity. The electric oven was showcased at the Chicago World's Fair a year later as part of a modern, fully electrified, and well-equipped kitchen. Unlike the gas stove, the electric stove was slow to catch on in the beginning due to the unfamiliarity of this technology and the fact that many towns and cities did not have electricity. Nevertheless, by the 1930s, this technology was much more refined and most towns and cities had electricity. As a result, the electric stove began to replace the gas stove in more and more homes and is commonplace today in many parts of the world.

Another Canadian invention that eventually became ubiquitous globally is the paint roller. This invention is said by many to have revolutionized the home renovation industry, because it is so easy to use as well as convenient and inexpensive. It was invented by Norman Breakey in 1940. Like many other Canadian inventors, Breakey did not have the funds to capitalize on his invention, or to ward off imitations by other inventors and disagreements with other patent contenders, especially Richard Adams in the United States, whose patent was granted subsequent to Breakey's. As John Melady points out in his book, *Breakthrough! Canada's Greatest Inventions and Innovations*:

> The paint roller continues to sell successfully in virtually every country around the world. It is available in the most humble hardware and home-furnishing store, and it is a staple product in every "big box" outlet as well. A cursory inquiry at one of our major dealerships indicated that four thousand paint rollers were purchased there in a single year. This was one store here in Canada. Think of how many rollers passed over the counters of all the places where the product is sold. Norman Breakey would surely be amazed.[6]

A third noteworthy invention is the transmission electron microscope. Working in close cooperation with three of his graduate students, James Hillier, Albert Prebus, and Cecil Hall, Eli Franklin Burton, Director of the McLennan Physics Laboratory at the University of Toronto, was involved in the creation and successful testing of the first practical transmission electron microscope (or "TEM") in North America in 1938. The significance of this invention, both for Canada and the world, should not be underestimated. The invention of the electron microscope by Ernst Ruska in Germany in 1931, and then the *transmission* electron microscope by Burton, Hillier, Prebus, and Hall in 1938, made it possible for scientists to see objects in incredible detail that only appeared as fuzzy shapes and shadows even in the most powerful optical microscopes. Using a beam of electrons, the TEM can magnify objects up to two

---

6. John Melady, *Breakthrough! Canada's Greatest Inventions and Innovations* (Toronto: Dundurn, 2013), p. 94.

million times. Suddenly, scientists were able to see things like viruses that had not been seen before.

The TEM has proved to be a major tool in biological and medical research, as well as in industry where it has been used to examine such substances as wood fibres, asphalt, textiles, plastics, dyes, inks, paints, metal surfaces, and many more.

While not all the creative achievements by Canadians in natural resources, agriculture, and industry have been included in this chapter, enough have been discussed to reveal why creativity in this area has been and remains of fundamental importance to people and countries in other parts of the world and the world as a whole. While Canada and Canadians have profited immensely from this creativity and have been the principal beneficiaries of it, so has the world at large, from the extraction and export of natural resources and basic staples to the creation and production of many agricultural, industrial, and technological products, devices, implements, and processes. Canada and the world as a whole are much better off today as a result of this.

Chapter Four
Medicine, Health Care, and Compassion

Canadians are often depicted as caring, sharing, and compassionate people. Whether they embody these qualities to a greater extent than other people in the world is impossible to say. What it is possible to say is that Canadians have manifested characteristics like this from the beginning of their history to the present day.

These qualities are most evident in the medical and health care systems that Canadians have evolved over the centuries. But they are also evident in the responses they have made and are making to many humanitarian crises in other parts of the world. These systems and responses manifest a great deal of creativity in numerous ways, since they are intimately connected to the problems that Canadians have had to confront and overcome over the course of their history.

Medicine and health care were still in their infancy in what would become Canada and in other parts of the world in the seventeenth and eighteenth centuries. Little was known about the nature and causes of most types of illnesses and diseases, especially in terms of how they were transmitted and could be dealt with effectively. As a result, epidemics and diseases such as diphtheria, measles, cholera, typhus, smallpox, and chicken pox were rampant. They reduced life expectancy in Canada substantially—usually to under 40 years—and caused the death of more than 50 percent of all newborn infants.

Although the situation had improved significantly by the first half of the nineteenth century, especially in cities where there were hospitals and some universities were providing medical courses and degree programs, effective medical practices were still lacking when William Osler arrived on the scene in the middle of the nineteenth century. He was born in Bond Head, Ontario in 1849 and educated at the Toronto Medical School and McGill University before heading off to Europe to undertake post-graduate studies. Following this, he taught at McGill University for several years before becoming a professor of medicine at the University of Pennsylvania in 1884.

In 1889, Osler was appointed first Physician-in-Chief at Johns Hopkins Hospital in Baltimore, which he helped to create with three other colleagues. He also played a seminal role in founding the Johns Hopkins School of Medicine, which became well known shortly after its creation as one of the most advanced medical schools in the world due to the innovative changes that Osler and others introduced there to medical practices, procedures, and policies. These changes were focused largely on improving the education of doctors and the training of medical students, primarily by placing a much higher priority on patients, patient care, and the cultivation of what Osler called "bedside manner." This necessitated the establishment of teaching hospitals where it was possible and necessary for students to spend a great deal of time working in hospital wards helping patients and doctors, rather than spending the bulk of their time in lecture halls and libraries. In an article Osler wrote on this matter called "Books and Men," he said, "He who studies medicine without books sails an uncharted sea, but he who studies medicine without patients does not go to sea at all." This is why one of Osler's favourite sayings was, "Listen to your patients, they are telling you the diagnosis."

These ideas spread rapidly throughout the United States,

Canada, Europe, and other parts of the world due to Osler's inventive teaching techniques as well as the enthusiastic response of his students and medical colleagues. This is particularly true for his students—especially a Dutch student intern named P. K. Pel—who spread the word about Osler and his practices to many medical schools throughout the world. In fact, Osler's methods and techniques have become a standard feature in virtually all hospitals and medical schools in the world today, regardless of where they are located. As a result of this, medical residencies involving third- and fourth-year students accompanying a chief physician or surgeon on his or her rounds, and doctors and students living together in the same administrative quarters in hospitals for significant periods of time, are commonplace practices in most if not all parts of the world today.

As these medical practices and procedures spread throughout the world, Osler's reputation grew rapidly. He became known as "the father of modern medicine" and "one of the greatest diagnosticians to ever wield a stethoscope." He spent many of his later years at Oxford University in England as Regius Professor of Medicine.

This is how Douglas Waug, a medical doctor, describes the pioneering contributions made by Osler:

> Renowned for his great personal charm, prodigious energy and outstanding ability, Sir William Osler had a profound impact on the field of medicine. He was a teacher who demonstrated a powerful affection for his students and a life-long love of books; he was a prolific writer of over 1,500 scientific publications; and he was a physician highly respected for his meticulous observation and devout scholarship. But most importantly, William

Osler was one of the innovators who propelled nineteenth century medicine into the modern era....

It is commonly said that William Osler was the greatest physician of the past century. As a clinician-consultant, teacher, medical educator, historian, classicist and biographer, he greatly influenced succeeding generations, and the effects of his medical philosophy are still visible today.[7]

Although Osler's medical interests ranged far and wide and were extremely diverse, he was particularly interested in different types of heart, lung, and blood diseases. These interests are often cited, along with his many other accomplishments, as playing a key role in the creation of the Royal College of Physicians and Surgeons in Canada in 1929. His popular book, *The Principles and Practice of Medicine*, published in 1892, was *the* standard textbook in medical circles for more than forty years.

Small wonder Osler's name is associated with outstanding medical procedures and practices in all parts of the world. According to Dr. Charles G. Roland, an accomplished Canadian doctor, there are many reasons why Sir William Osler is known as "one of the greatest medical masters the world has ever known." Not only was he extremely creative, but he was also caring, sharing, and compassionate—"a person who put the interests of his patients and his students ahead of his own interests." As Roland said in the following tribute to Osler:

> His importance derives largely from five inter-related areas: his contributions to medical knowledge through

---

[7]. Ted Grant, *Doctors' Work: The Legacy of William Osler* (Altona, Man.: Firefly Books, Inc., 2003) p. 15 and p. 49.

clinical and pathological research; his activities as a gifted educator; his ability to stimulate and inspire students who later became leaders of the medical profession in North America; his love of books and his unselfish support of medical libraries; and—perhaps most important to students and practitioners today—his quite unconscious role as an exemplar of integrity, humanity, kindliness, and professional honesty.[8]

While Osler is one of the best-known medical doctors Canada has ever produced, he was not the only Canadian doctor to make creative contributions to the advancement of medicine and health care that have had a powerful effect on the world. Canadians were back at the forefront of medicine once again in the early decades of the twentieth century for some equally creative and timely contributions, in large part because of the intense competition between Toronto and Montreal, the two main medical centres in Canada and some might say the world at that time.

In Toronto, there was the pioneering work of Frederick G. Banting, Charles H. Best, J. J. R. Macleod, and J. B. Collip. In 1921–22, they discovered insulin, the hormone in the body that regulates the level of glucose in the bloodstream and is produced by the pancreas. When insulin is not produced in sufficient quantities, the result is diabetes, an affliction that affects millions of people throughout the world.

While each of these medical pioneers had his own particular idiosyncrasies, they are best exemplified by Banting, who was the driving force behind the discovery of insulin along with his

---

8. Charles G. Roland, "The History of Canadian Health Care," in Charles J. Humber, editor-in-chief, *Canada: From Sea to Sea* (Mississauga, Ont..: The Loyalist Press Limited, 1986), p. 471.

colleague and close friend Charles Best. After reading an article in a medical journal, Banting scribbled down some notes for a research study he wanted to undertake that was aimed at understanding the production and function of insulin in the body. He received support for this study in 1921 and immediately set to work on this study under the direction of Macleod with assistance from Best and Collip. While the first experiments were crude and unsuccessful, Banting and the group persevered until they finally succeeded in discovering insulin a year or so later.

Through experiments on dogs and cattle, an economical way was found to produce insulin that could be injected into people suffering from this dreaded disease, which includes type one diabetes, which is acquired at birth and is usually deadly if not treated effectively, and type two diabetes, which develops later in life. Type two diabetes is often described as "the silent killer" because many of the early symptoms do not manifest themselves until it is too late, when far more serious symptoms, including foot problems, ulcers, weight loss, kidney failure, blindness, heart attacks, and strokes make themselves apparent.

The creative breakthrough made by the Canadian team and especially Banting instantly catapulted the group into world prominence, particularly during and after 1922 when they saved the lives of Leonard Thompson, a 14-year-old boy suffering from type one diabetes, and Elizabeth Hughes, a 15-year-old girl also suffering from this disease. What made saving the life of Elizabeth Hughes so internationally noteworthy was the fact that she was the daughter of the famous American politician, Charles Evans Hughes, who was a governor, secretary of state, and chief justice of the U.S. Supreme Court.

Word spread rapidly after Hughes was saved from death that

a group of Canadian doctors had discovered a cure for diabetes. The problem was that the supply of insulin they possessed at this time was woefully inadequate to meet the demand. As a result, thousands of people had to be turned away in disappointment as Banting, Best, and Collip laboured night and day to find a practical way of increasing the supply of insulin.

What makes the discovery of insulin so remarkable is the fact that, once the supply of insulin was increased, millions of lives were saved, with the potential to save millions more in the future. Diabetes is one of the fastest growing diseases in the world today and is threatening to escalate out of control due to a variety of factors, such as obesity, excessive consumption of products high in sugar, lack of physical exercise, and others. A recent study by Stanford University, for instance, indicated that 511 million adults around the world are expected to have type two diabetes by 2030, up from 406 million in 2018, with more than half of these new cases in China, India, and the United States. According to the study, "these estimates suggest that current levels of insulin access are highly inadequate compared to projected need, particularly in Africa and Asia, and more efforts should be devoted to overcoming this looming health challenge."

Since it was first discovered, numerous improvements have been made in the quantity and quality of insulin, as well as in the way it is injected into the body. Initially, insulin was extracted from the pancreas of animals, most often pigs and cattle. However, major breakthroughs have been recorded in recent years in the creation and production of this drug from human sources, as well as in the production of what are called "insulin analogues," man-made compounds that are similar to natural insulin but faster to take effect and with longer-lasting results.

Major breakthroughs have also occurred in the early detection of diabetes, as well as in making it possible for diabetics to check their own blood glucose levels on a regular basis and monitor and manage their condition much more effectively. Whereas in the formative years, insulin was injected into the body through syringes that had reusable needles that often became blunted or infected, today small plastic insulin syringes are used that can be discarded after a single use.

It is also much easier for diabetics to manage their own condition today through the use of "insulin pumps." These pumps are designed to deliver fast-acting insulin 24 hours a day through a catheter placed under the skin. This technique enables people suffering from this disease to adjust their insulin levels on a more continuous basis, as well as in ways that suit their lifestyles and daily routines more effectively. A great deal of research is also being undertaken into the possibility of stem cell transplants designed to increase the production of insulin in the pancreas, rather than having to rely on perpetual injections. These improvements, and others, are enhancing prospects for the future, building on the initial discovery of insulin, which has been called "one of the greatest discoveries in the world in the twentieth century." In terms of saving and improving lives, it is certainly at or near the top of the list. Not surprisingly, two of the members of the original group of scientists, Banting and Macleod, were awarded the Nobel prize for this incredible achievement in 1923. Banting was the first medical doctor born in Canada to receive this prestigious award.

Just as Toronto was a hot bed of medical research and creativity, so too was Montreal. In the late nineteenth century, a group of medical doctors and practitioners created the Montreal General Hospital, established the first real medical school of its type in Canada at McGill University, and gained control of the medical licensing board of Quebec. Medical standards improved rapidly after that, since many regulations were put in place to ensure that patients got the best medical treatment and health care available at that time.

By the early decades of the twentieth century, Montreal boasted many medical experts destined to have outstanding careers in medicine and attract large international followings for their pioneering work. One of the most important of these was Wilder Graves Penfield, one of Canada's and the world's most inventive and influential medical minds of the first half of the twentieth century.

Penfield was born in the United States in 1891 and trained in Europe as well as in the United States under Harvey Cushing, Charles Sherrington, and Sir William Osler before graduating in medicine from Johns Hopkins University in 1918. He came to Canada in the 1930s as a co-founder (with William Cone) as well as the first director of the Montreal Neurological Institute, which was funded through a major grant from the Rockefeller Foundation in the United States and also supported by the government of Quebec, city of Montreal, and private donors. Penfield also worked at McGill University and the Royal Victoria Hospital in Montreal. Under Penfield's skillful leadership, the Montreal Neurological Institute quickly became well known throughout the world for its innovative research, publications, teaching, and treatment of diseases of the brain and nervous system.

As a neurosurgeon, Penfield was interested in how the human brain functions under normal and abnormal conditions and the way it affects different parts of the body. He was especially interested in epilepsy, as well as how the brain and cortical homunculus function with respect to speech, memory, motor skills, and sensory stimulations of one kind or another. He is credited with focusing attention on the two hemispheres of the brain and the many different skills, abilities, and characteristics the right and left hemispheres possess. His research and writing quickly led to many studies in Canada and other parts of the world on the functioning of the brain and its two hemispheres, something that Penfield felt was the key to understanding human beings in general, as evidenced by his oft-quoted remark, "The problem of neurology is to understand man himself."

Penfield greatly expanded the methods and techniques used in brain surgery. He was one of the first doctors to map the brain and its various regions and components in great detail. He also showed how the sensory, motor, parietal, and temporal cortices of the brain are linked to the various limbs and organs of the body. Keenly interested in the relationship between the brain and the mind, Penfield also provided insights into how dreams, hallucinations, and the feeling of déjà vu occur and how memory works, as well as where the brain's speech and speech comprehension centres are located. He often used a pencil-tip sized electrode with a mild current to gently stimulate different areas of his patients' brains in order to determine what they experienced in terms of sight, smell, feelings, sounds, and so forth.

In the years to follow, Penfield's pioneering work laid the groundwork for many intensive studies of the brain and its various functions and capacities in Canada and elsewhere in the world. These studies have enhanced Penfield's international reputation as a seminal contributor to brain research, a field that is so crucial, timely, and important today. Penfield was also a co-founder of the Vanier Institute of the Family, which further enhanced his reputation as an outstanding researcher and medical pioneer. For these and other creative achievements with international consequences, Penfield was honoured by Google on January 26, 2018, which would have been his 127th birthday.

One doctor who followed in Penfield's footsteps and made her own remarkable contributions to neurological activities and functions was Brenda Milner. She came to Montreal from Manchester, England to work with Penfield and then served as director of neuropsychology at the Institute in the 1980s, where she worked until she was well into her nineties. She has often been described as "the founder of neuropsychology" for the pioneering work she conducted on brain development and especially memory in general and memory loss and retention in particular. Just as Penfield's pioneering work on the brain paved the way for Milner's work in neuropsychology and memory and its loss and retention, so Milner's work paved the way for the next generation of Canadian researchers to do exciting work in the rapidly expanding field of neuropsychology.

One of these researchers is Lola Cuddy, who created one of the first proto-musical psychology laboratories in the world at Queen's University in Kingston, Ontario to study the relationship between music and the brain. As it is for music, so it is for many other arts and their relationship with the brain. Growing interest in this field resulted recently in the creation of the International Arts+Mind Lab at the Johns Hopkins Brain Sciences Institute as a multidisciplinary research-to-practice initiative that is accelerating the growth of neuroaesthetics through impact-based studies, research, and collaboration. How fitting is this recent development, given the fact that many important medical innovations got their start at Johns Hopkins a hundred years ago under Sir William Osler!

As a result of Penfield's work and that of many others, the medical profession has a much better understanding today of such complex matters as motor, sensory, speech, and memory skills, epilepsy, consciousness, creativity, rationality, linguistic capabilities, and a great deal else. What makes the numerous articles, handbooks, monographs, and books Penfield wrote on these and other subjects so important is not only the accolades and awards they received at the time, but the foundation they laid for contemporary developments. Just as Osler paved the way for crucial improvements in patient care, so Penfield paved the way for comparable developments in brain research, one of the most active and important fields in contemporary medical research today.

Indeed, Penfield's creative work also foreshadowed the rapidly escalating interest in neuroplasticity evident at the present time. This field is concerned with how the brain changes over time in response to a variety of behavioural, environmental, physical, social, and educational stimuli and factors. Whereas it was once thought that the brain was largely static and didn't change a great deal over time—much like the motor in a car—it is now thought that the brain is dynamic and is in a constant state of evolution and flux due to its inherent malleability and flexibility. This is providing a ray of hope for many conditions that were thought to be unchangeable in the past but which now can be changed and changed significantly. This new approach is apparent in the research and writing of many doctors and researchers working in this field today, such as Norman Doidge, a Canadian doctor who has written a number of books on the subject of neuroplasticity, most notably *The Brain that Changes Itself* and *The Brain's Way of Healing*, as well as neurophilosophers such as Patricia (Smith) Churchland and Paul Montgomery Churchland, two Canadians who have written at great length about the connections between the brain and the mind in the context of the philosophy of science.

As a result of these developments and the contributions of other highly creative Canadian medical researchers and doctors such as Freda Miller, Doreen Kimura, Peter St. George-Hyslop, Ronald Melzack, Gillian Einstein, and others, "Canada is considered one of the birthplaces of neuroscience and continues to account for a large portion of the research being conducted in this field," according to the Brain Canada Foundation. The Foundation goes on to say: "Recent bibliometric study has found that Canada is the most specialized country in the world in neuroscience, and that it continues to rank among the top five countries in the world in terms of research output and research impact in this field. In recent years, brain research has expanded beyond a singular focus on neuroscience, to a more multidisciplinary approach, with different fields converging. Canada has strengths in areas key to advancing brain research, including genomics, proteomics, cell biology, imaging, big data, and especially Artificial Intelligence."

Yet another highly creative medical researcher and pioneer in Montreal during much of the twentieth century was Hans Selye, who specialized in, and popularized, research on "biological stress," especially in humans. Educated in Prague, Paris, and Rome, he joined McGill University in 1932 and became the first director of the Institute of Experimental Medicine and Surgery at the University of Montreal in 1945, where he remained until his retirement in 1976. He also created the International Institute of Stress in 1975, and later, in conjunction with eight Nobel laureates, founded the Canadian Institute of Stress.

Selye contended that stress plays a crucial role in the development of many chronic illnesses and diseases, such as hypertension, peptic ulceration, renal disease, arthritis, asthma, cancer, and so forth. He identified three major phases in the stress process: alarm; resistance and adaptation; and exhaustion. Inability to deal with stress in general, and with these three phases in particular, leads to various "diseases of adaptation," according to Selye. As a prolific author with more than a thousand articles to his credit, his most influential article was "A Syndrome Produced by Diverse Nocuous Agents," which was published in 1936 in the *Journal of Neuropsychiatry and Clinical Neurosciences*. He also authored a number of books dealing with endocrinology, stress, and disease, such as the *Encyclopedia of Endocrinology, The Stress of Life, Stress without Distress,* and *The Stress of My Life*. His ideas and theories on these matters were widely publicized throughout the world and were often reported in national newspapers, magazines, and clinical journals. *Time* magazine, for example, ran a series of cover stories on the multiple manifestations of stress in the lives of Americans which was based largely on Selye's beliefs on this subject. While some people thought Selye should have received a Nobel prize for his creative achievements and efforts in this area, he was likely too controversial and provocative to receive this award.

Indeed, this is still true of Selye and his reputation today. Assessments of his contributions and achievements are still being hotly debated and are very controversial at present, with some experts downplaying his influences on contemporary medical developments and practices and others applauding and promoting them. One of Selye's own students, Roger Guillemin, probably summed this ambivalence up best when he said: "Selye ... was the source of many ideas which, whether accepted or, more often, challenged, whether confirmed as such or eventually profoundly modified, were at the roots of modern neuroendocrinology."

And this is the point. Regardless of which side of the debate medical experts come down on, there is one matter on which they all seem to agree. Selye made a valuable contribution to stress and adaptation theories and practices as well as to modern neuroendocrinology by focusing on these subjects in medical circles, journals, institutions, and public appearances. What makes this so important is the fact that stress and actual and potential adaptations to it have become prominent features and factors in the lives of billions of people throughout the world today. Given the state of the world at present and prospects for the future, it is highly likely that in the years ahead, more and more medical researchers and doctors will turn their attention to the causes, consequences, and treatment of stress, and with this, inevitably to the writings and findings of Selye on this subject.

About the same time that Selye and Penfield were conducting their medical research in Montreal, a number of Canadian medical doctors were travelling to China. They were sent there by various medical, religious, governmental, and non-governmental organizations to assist China and the Chinese people in their devastating war with Japan, one of the worst famines to hit the Middle Kingdom in centuries, and a civil war that ultimately brought an end to the neo-colonial phase of Christianity in China.

One of these doctors, Norman Bethune, was a well-known surgeon, inventor, activist, and humanitarian. He spent the bulk of his life helping others, primarily in Spain and China. He interrupted his medical studies at the University of Toronto for several years to become a labourer-teacher at Frontier College and then acted as a stretcher-bearer during the First World War. After spending a number of years in the Royal Navy, he contracted pulmonary tuberculosis and devoted the next few years of his life to helping tuberculosis victims and undertaking thoracic surgery and research in Montreal at the Royal Victoria Hospital and later the Hôpital du Sacré-Coeur in Carterville, Quebec. He invented many medical procedures and surgical instruments during this time, and wrote numerous papers describing innovative thoracic techniques that were helpful in different parts of the world.

After visiting the Soviet Union in 1935, Bethune joined the Communist Party and immediately took up the communist cause, first in the civil war in Spain where he created the world's first mobile blood bank and transfusion service—something still used in many parts of the world today when required—and later in China during the Second Sino-Japanese War. He laboured intensively in China for the next two years of his life on behalf of the Chinese people in general and the Chinese Communist Party in particular, not only as a surgeon and inventor but also as an activist, teacher, and friend. He died in 1938 from an accidental cut he received while performing an operation in China.

Bethune was a household name and hero in China. This resulted not only from his medical contributions there, but also from the fact that he was immortalized in a famous essay by Mao Zedong—*In Memory of Norman Bethune*—which extolled all communists, and especially all Chinese communists, to emulate Bethune's virtues of humanitarianism, internationalism, sense of responsibility, commitment to good health, and dedication to helping others. This did more than anything else to create the bond that existed between Canada and China for over 70 years. It was a bond that yielded countless benefits for both countries, not only in medicine, but also in the arts, sciences, humanities, industry, resources, and virtually every other field, despite the fact that this bond has been strained over the last few years due to the arrest of the Huawei executive Meng Wanzhou in Canada, the jailing of two Canadians in China, and the canola seed conflict discussed earlier. This bond was attributable largely to the impact Norman Bethune had on China and the Chinese people, which explains why there are many statues of Bethune in different parts of China today.

Bethune was not the only Canadian doctor to have a "China connection," or to undertake creative medical work that became known and utilized throughout the world. Another was Harold Johns.

Johns was born in a Buddhist Temple near Chandu, China in 1915. His father was a missionary who also taught mathematics at West China University. After the family returned to Canada in 1927, Johns went to university where he received a B.A. in physics from McMaster University in 1936 and a Ph.D. from the University of Toronto in 1939. Like a number of other famous Canadian inventors and scientists, including Nobel prize-winners Gerhard Herzberg and Henry Taub, Johns spent a good deal of time at the University of Saskatchewan where he taught from 1945 to 1956. He became interested in biophysics, and especially the potential cobalt possesses to treat cancer.

What especially fascinated Johns about cobalt was its ability to generate a much more powerful radiation than radium, which was used extensively at that time to treat cancer due to its ability to be focused on a single point in the body and literally "bombard a tumour." After years of research on radioactive cobalt and the use of reactors, Johns invented the cobalt-60 bomb. It was installed in the University Hospital in Saskatoon and Victoria Hospital in London, Ontario in 1951 by Atomic Energy of Canada.

Victoria Hospital was the first institution in the world to utilize the cobalt-60 bomb on October 27, 1951. The new technology made it possible to treat deep tumours in the body for the first time, such as those in the bladder, cervix, and lungs. The "cure rate" for cervical cancer, for instance, jumped from 25 percent to 75 percent soon after the development and utilization of the cobalt-60 bomb. Its use drastically reduced the amount of time required for cancer treatments, from more than an hour to roughly five minutes, as well as yielding some exceptional results in patients' recoveries. In 1953, Johns published *The Physics of Radiation*, and in 1956 became Head of the Physics Division at the Canadian Cancer Institute as well as Professor of Medical Biophysics at the University of Toronto.

According to the London Health Sciences Centre, approximately 35 million cancer patients throughout the world had benefitted from this ground-breaking technology by 2011. It is still in use in approximately 70 percent of the world's cancer cases treated by radiation today, despite the fact that many other advances and treatments are now available as well. This is due primarily to the cost effectiveness of the cobalt-60 technology, as well as its reliability and simplicity of use.

Johns' work in this area was also helpful in projecting Canada into a prominent position as a world leader in nuclear medicine and in developing peaceful and innovative applications of nuclear technology through the country's expertise in physics, metallurgy, chemistry, biology, and engineering. Much of this occurred as a result of the establishment of the Canadian

Nuclear Laboratories, especially the Chalk River Laboratories, as well as the crucial role Canada played and is still playing in the development of medical isotopes as well as being the largest supplier in the world of molybdenum-90, which is generally seen as the "workhorse" and most commonly used isotope in nuclear medicine today. Interesting, radioactive isotopes are now used as a diagnostic tool, as a treatment for cancer through the gamma rays emitted from the isotopes directly to the source of a tumor, and in turning research findings into practical products and pharmaceuticals in many parts of the world.

Unlike Harold Johns, who was born in China but did most of his creative work in Canada, Daniel David Palmer was born in Canada but did much of his creative work in the United States. Palmer was born in 1845 and grew up in Port Perry, Ontario. When the family business failed, his parents moved to the United States to look for work. After living with his grandparents in Ontario until he was in his early twenties, Palmer decided to leave his grandparents and join his parents in the United States.

Palmer was a beekeeper, school teacher, and grocery store owner before he found his calling and principal passion in life. It was "chiropractic," which he created in Davenport, Iowa in 1895 where he was living at that time. He spent many years as a "magnetic healer" there before he created chiropractic and established the basic principles and practices that underlie and govern this field today. Chiropractic was generally seen at that time and for a long time thereafter as an alternative medical field concerned with the diagnosis and treatment of "unverified" disorders of the musculoskeletal system in general and the spine in particular. It is predicated on making adjustments to the vertebrae in the spine so that the spine can function properly, thus resulting in better health.

Palmer's reputation as the founder of chiropractic grew rapidly after he cured a janitor, Harvey Lillard, who had been deaf for 17 years after he exerted himself too strenuously in a cramped, stooped position and felt something give in his back. An examination by Palmer revealed that the vertebrae were not in their proper places. Believing that Lillard's hearing would be restored if the vertebrae were repositioned, Palmer snapped the vertebrae back into their normal position, which ended up restoring Lillard's hearing. This provided the impetus that was necessary for chiropractic to achieve some traction as an important field of medical activity.

Describing himself as "the one who discovered the basic principles of chiropractic, developed its philosophy, originated and founded the science and art of correcting abnormal functions by hand adjusting, using the vertebral processes as levers," Palmer published the first book on chiropractic, which was titled *The Text-Book of the Science, Art, and Philosophy of Chiropractic, or The Chiropractor's Adjuster*. It was published by the Portland Printing House Company in 1910 and republished in 1966. Palmer also created a School of Chiropractic in Davenport. Interestingly, there is a statue of Palmer in a park by Lake Scugog in Port Perry that recognizes this remarkable individual's highly creative achievements in this area. At the bottom of the statue is a revealing statement by Palmer that reads, "I have never considered it beneath my dignity to do anything to relieve human suffering."

Despite Palmer's success in founding chiropractic, acceptance of this field as a legitimate area of medical theory and practice was fraught with difficulties from the outset. Much like acupuncture, there was strong resistance to chiropractic in the western world for many years and it took an incredible amount of time, effort, and advocacy before the western medical community recognized it in a formal sense. It also took strenuous efforts on the part of three generations of Palmers: Daniel David, "the founder" of chiropractic; Bartlett Joshua, his son, who was often described as "the developer" of chiropractic for the pioneering role he played in promoting the actual practice of chiropractic; and Bartlett Joshua's son David Daniel, who has been called "the educator" for the pivotal role he played in building the Palmer College of Chiropractic in Davenport and similar institutions in other parts of the United States and elsewhere in the world.

As a result of these developments, and others, chiropractic is now a much more accepted and established field of medical theory and practice, although it still has its naysayers and deniers as well as its practitioners, advocates, and promoters. It is now well established in the United States, Canada, England, and Australia, and is making rapid progress in many other parts of the world. Moreover, there is now a World Federation of Chiropractic, which was founded in 1987 and has national associations representing 88 countries in the world as a non-governmental organization affiliated with the World Health Organization. The headquarters of this Federation are in Toronto, a fitting tribute to the fact that its founder and first practitioner was born in Canada and spent more than two decades here.

Like Daniel David Palmer, Elizabeth McMaster was a real medical and health-care pioneer. She played a key role in the creation of the first hospital for sick children in North America, and only the second in the world after Great Britain. Concerned about the fact that nearly half of all deaths recorded in 1875 were children under the age of 10, McMaster and a group of dedicated women rented an 11-room house on Avenue Road in downtown Toronto and declared open a "hospital for sick children."

While it contained only six iron cots initially, this hospital has grown to become the world-famous Hospital for Sick Children—now often known as "Sick Kids"—with locations on Elizabeth Street, College Street, and University Avenue in the central part of downtown Toronto. It is renowned throughout the world for its creativity and innovation. While very few Canadians are aware of it, Pablum—one of the world's most popular baby foods at one time—was created by doctors and medical specialists connected with this hospital, including pediatricians Frederick Tisdall, Theordore Drake, and Alan Brown, nutritionist Ruth Herbert, and chemists Mead Johnson and Harry Engel. Pablum's popularity was predicated on its ease of preparation, as well as its ability to prevent rickets—a crippling childhood disease—by ensuring that children had enough vitamin D in their diets. It was the first precooked and thoroughly dried baby food in the world, which made it popular in all parts of the world because it was inexpensive and could be mixed with water or milk to produce a smooth, nutritional, and digestible paste. As the second largest pediatric research hospital in the world today, Sick Kids has also played a major role in pioneering pasteurization, the iron lung for polio, the Mustard procedure for "blue babies," discovery of the gene for cystic fibrosis by a team of doctors and researchers headed by Lap-Chee Tsui, and many other advances.

Canada's creative medical tradition was also carried on in an excellent fashion by Wilfred Bigelow. He was born in Brandon, Manitoba to parents who were both involved in the medical profession. His father, Wilfred Abram Bigelow, was the founder of the first private medical clinic in Canada, and his mother, Grace Ann Gordon, was a nurse.

Following his graduation from the University of Toronto in 1938, Bigelow served as a captain in the Royal Canadian Medical Army Corps during the Second World War and conducted many surgical procedures on the battlefield. After the war, he became a member of the surgical team at the Toronto General Hospital in 1947 after spending a year at the Johns Hopkins Medical School. He joined the Department of Surgery at the University of Toronto in 1948.

One of Bigelow's most important discoveries was made in 1950 when he realized how it was possible to lower the oxygen requirements in the body by reducing the body's basic temperature to a point where open-heart surgery was possible. The first human application of Bigelow's hypothermia research in heart surgery occurred in 1952. Interestingly, he had already pioneered another major development in the treatment of heart disease—the creation and refinement of the artificial heart pacemaker, a device that is placed in the chest or abdomen to help control abnormal heart rhythms and uses electrical impulses to prompt the heart to beat at a normal rate. Bigelow developed this device initially in 1949, working in close cooperation with Dr. John Callaghan and later, in a much more advanced stage, drawing on the engineering expertise of Dr. John Hopps. The pacemaker and its potential were explained at the annual meeting of the American College of Surgeons in Boston on October 23, 1950. There were over 9,000 doctors at this meeting from many different parts of the world. The news was greeted with a great deal of enthusiasm and spread rapidly. Although the first pacemaker was very large and could not be inserted into the body, this changed dramatically when subsequent versions incorporated transistors and miniature batteries.

In combination, the pacemaker and Bigelow's research on hyperthermia had a profound effect on heart surgery and consequently on the lives of millions of people with heart disease around the world. Much of Bigelow's knowledge and expertise in this area is set out in his book *Cold Hearts*. He was inducted into the Canadian Medical Hall of Fame in 1981 with a citation that reads, "One of the most distinguished surgeons Canada has ever produced and [someone who] stands among the world's titans in medicine." His achievements have proven helpful in spawning an enormous amount of research into heart disease in general and major improvements in the pacemaker in particular by doctors in Canada such as Ray Chiu, Tirone David, and Wilbert Keon, as well as by those in other parts of the world. Pacemakers are now an indispensable tool of modern medicine.

Canadian doctors were once again in the news in this field because of the recent work of John Webb and others in Vancouver, who achieved another major breakthrough in heart surgery. It involves entering the patient's body through the foot rather than the chest and sending new valves up to the heart through the veins. This is particularly helpful for patients who have had a number of major heart operations and cannot endure any more. This procedure is being studied carefully by medical practitioners elsewhere and has been performed successfully across North America and other parts of the world.

Canada's creative medical contributions to the world were enhanced even more in recent years by two Canadian medical giants, Ernest McCulloch and James Till. These two individuals conducted a great deal of highly original research over many decades on stem cells and regenerative medicine at the Princess

Margaret Hospital in Toronto and other medical institutions and facilities in Ontario. They established the existence of stem cells nearly 40 years ago and are deemed by many in the medical field to be "the fathers of stem cell research" in Canada and throughout the world. They were inducted into the Canadian Medical Hall of Fame in 2006.

Working at the Ontario Cancer Institute in the early 1960s, McCulloch and Till published a ground-breaking paper in 1961 on what are known as "colony-forming cells." This paper laid the foundations for stem cell research, setting out many of the basic concepts, principles, theories, and terms used in this field today. A specific example involves bone marrow transplantation for cancer patients, in which Till and McCulloch established the theoretical underpinnings for therapies some 40 years ago, thereby saving countless lives.

While the work of these accomplished doctors and researchers has been highly controversial because of the aversion to stem cell research in certain medical, political, and religious quarters, this research is having a profound effect on current medical practices and is regarded as "the way to go in the future" by many in the medical profession. Creative work in this field has placed Canada in the forefront of stem cell research throughout the world over the last few decades, with major work being carried out on cancer stem cells and blood stem cells, as well as on human embryonic stem cells by Mickie Bhatia at McMaster University and stem cells that defy rejection and disease by Derrick Rossi, a Toronto-born researcher working at Harvard University. As well, the McEwan Centre for Regenerative Medicine in Toronto and the Stem Cell Network in Ottawa have played an important role in research in this field.

Toronto scientists achieved another medical milestone in stem cell research recently when Andras Nagy and his team at Mount Sinai Hospital in Toronto, in conjunction with Keisuke Kaji of the Medical Research Council Centre for Regenerative Medicine at the University of Edinburgh, became the first researchers to reprogram adult human cells into embryonic-like cells without using potentially dangerous viruses that can cause cancer. This breakthrough is expected to have a major impact in overcoming one of the biggest obstacles to using the reprogramming technique in the development of new drugs, finding cures for many diseases, and creating personalized organs and tissues.

Another major contribution to the rapidly growing field of stem cell research has been made by Tak Wah Mak. Son of a businessman from southern China, he went to the University of Wisconsin in the late 1960s and then completed his Ph.D. at the University of Alberta in the early 1970s. Following this, he undertook a great deal of research at the Ontario Cancer Institute in Toronto. Most of his earlier research centred on increasing understanding of the biology of cells in normal and diseased settings and focusing specifically on what is known as "T-cell development, activation, and differentiation." In 1984, Mak co-discovered the T-cell receptor and the gene that produces it, a finding which is generally regarded as one of the major keys to the human immune system. There are many different types of T-cells, which makes this one of the most important areas of immunology because T-cells can act in negative as well as positive ways. In 2004, Mak was made Director of the Institute for Breast Cancer Research at the Princess Margaret Hospital in Toronto and continues to practise there today. Since that time, he has become deeply immersed in cancer research and is one of Canada's and the world's most cited medical doctors, researchers, and scientists in this field.

It is clear from the foregoing examples and others that could be provided—most notably the creation of Medicare by Tommy Douglas and others in the early 1960s that transformed health-care practices, policies, and funding and provided Canadians with an inclusive national health-care system—that there is a strong relationship between medicine and health care in Canada on the one hand and caring, sharing, and compassion on the other. In both these areas, Canadians have directed their talents and energies towards helping their fellow citizens. It would be a mistake to push this argument too far, or fail to recognize that many social, religious, voluntary, and governmental organizations and agencies have also had a profound effect. However, this connection helps to explain why Canada has made so many creative contributions to medicine and health care.

The multifarious benefits that have flowed from these creative contributions over the years have not been confined to Canada and Canadians, but have had a powerful effect on other people and countries as well as the world as a whole. Whether it is the advent of modern medicine, the discovery of insulin, research on the nature and functioning of the brain, the mind, and the heart, dealing with stress, the treatment of cancer, the creation of mobile blood banks, the invention of Pablum, the artificial pacemaker, stem cell research, or all the other breakthroughs made by Canadians in medicine and health care, the world is much better off today as a result of Canada's creative contributions in these areas.

What is true with respect to medicine and health care is also true of the many creative contributions Canada and Canadians have made to humanitarian causes and concerns throughout the world. While there are many examples of this, some of the most obvious ones are the compassionate and timely responses the country and its citizenry have made to a number of pressing refugee problems, most notably the responses to refugees from Hungary, Southeast Asia, and Syria in the last half of the twentieth century and first two decades of the twenty-first century.

This started with the generosity Canadians demonstrated by taking in more than 37,000 Hungarian refugees after the abortive Hungarian Revolution in 1956. While many of these refugees shared certain values and customs with Canadians, what was most apparent was how much caring, sharing, and compassion Canadians displayed towards these refugees after they arrived in Canada and how effectively many of them were integrated into Canadian society.

An even more revealing example of this was the response by Canadians to the refugee crisis in Southeast Asia and especially Vietnam, Cambodia, and Laos in the late 1970s in the aftermath of the Vietnam War. Between 1975 and 1976, over 6,500 "boat people," as they were called, were admitted into Canada through a program that was devised and carried out by the government of Canada in conjunction with groups of Canadian citizens across the country. It was the first time the government implemented a highly creative program involving private and public sponsorships of refugees—the first of its type in the world—for which the country received the Nansen Medal from the United Nations. By the end of 1980, approximately 60,000 refugees from Southeast Asia had been accepted under this sponsorship program, with 26,000 sponsored by the federal government and 34,000 sponsored by Canadian citizens. By the end of the 1980s, more than 200,000 Vietnamese, Cambodian, and Laotian refugees had been settled in Canada in this way, a very large number compared to the size of the Canadian population.

This same collaborative model was employed once again during the heart-wrenching Syrian crisis when more than 50,000 Syrian refugees were accepted and settled in Canada as of January 2018 through the provision of food, clothing, accommodation, education, language training, employment opportunities, and a great deal else. While the program was not been without its difficulties, especially as a response to a crisis of this magnitude was expected quickly and pressure to accept, settle, and integrate refugees was very intense, it has nevertheless proved successful on the whole under extremely difficult circumstances. It is also reflective of the general outpouring of compassion and creativity that Canada and Canadians have demonstrated towards people who are in need of help.

To this must be added the generous response of Canada, Canadians, and Canadian governments to the terrorist attacks on the World Trade Center and the Pentagon on September 11, 2001. This response, dubbed Operation Yellow Ribbon, occurred especially when American airspace was closed to commercial aviation and roughly 33,000 passengers on 224 flights on route to the United States were suddenly sent to airports on Canada's east and west coasts in general and Gander, Halifax, and Vancouver in particular.

While caring, sharing, and compassion were demonstrated in all parts of the country in the reaction to this tragic situation and despicable event, it was particularly pronounced in Newfoundland and Labrador and most noticeable in Gander when 38 large passenger planes heading west over the Atlantic Ocean with some 6,700 weary, shaken, and exhausted passengers were suddenly diverted to "a safe harbour of welcome and warmth amid shock and chaos" at the airport in Gander, a town of roughly 10,000 people. The outpouring of love, affection, and compassion that followed this was incredible and unfathomable. Striking school bus drivers laid down their picket signs and drove passengers wherever they needed to go; pharmacists filled prescriptions without charge; shop owners declined payments for goods; passengers phoned loved ones in other parts of the world for free; the Gander Community Centre Arena became a giant walk-in refrigerator for donations of food and perishables; and masses of clothing suddenly appeared and seemingly from nowhere. Moreover, myriads of meals were prepared every day for people with many different culinary requirements and ethnic backgrounds; thousands of cots, beds, blankets, and pillows were provided on a voluntary basis in schools and community centres as well as in hotels and motels; and on and on it went until the passengers were able to carry on to their original destinations, homelands, and residences.

This incredible outpouring of love and compassion, which is still being celebrated and exhibited today, was subsequently made into an extremely popular musical, *Come from Away*, that was researched, written, and produced by two creative Canadians, Irene Sankoff and David Hein. It opened on Broadway to enthusiastic audiences and rave reviews, played at the Mirvish theatres in Toronto to sold-out houses for more than two years, was nominated for seven Tony Awards in the United States in 2017, won the award for Most Outstanding Production in the Musical Theatre Division at the Dora Mavor Moore Awards in Canada in 2018, and was named best new musical, choreography, sound, and most outstanding achievement in music at Britain's Olivier stage awards. It is still drawing huge crowds and appreciative audiences today with no end in sight. Little wonder Canadians are known throughout the world as being caring, sharing, and compassionate people.

The True North: Images of Canadian Creativity

Top left: Astronaut Chris Hadfield, spacewalking during the 2001 STS-100 space shuttle mission *(NASA)*. Top right: Astronaut David Saint-Jacques, who holds the record for longest single space flight by a Canadian at 204 days, aboard the International Space Station in 2019 *(NASA)*. Bottom left: Environmentalist David Suzuki *(Stephen Michael Barnett/Wikimedia Commons)*. Bottom right: General Roméo Dallaire *(St. Joseph's Health Care Foundation)*.

The True North: Images of Canadian Creativity

Top: Greenpeace has become one of the world's best-known environmental organizations *(RNW/Flickr)*.
Bottom: Author and activist Naomi Klein addresses an Occupy Wall Street rally in 2011 *(David Shankbone)*.

The True North: Images of Canadian Creativity

Left: Daniel David Palmer, the founder of chiropractic *(Public domain photo courtesy Roma Chiropractic)*

Below: Catriona Le May Doan and Rick Hansen at the opening ceremony of the 2010 Olympics in Vancouver *(Jeff Seifred).*

The True North: Images of Canadian Creativity

Above: Cirque du Soleil show in Vienna, Austria, in 1994 *(Clemens Pfeiffer, Wikimedia Commons)*.
Below: World-renowned Canadian pianist Oscar Peterson performing in Munich *(Hans Bernhard)*.

The True North: Images of Canadian Creativity

Left: Rapper, actor and songwriter Drake in 2017 at the Velvet Underground *(Anton Mak)*.

Below: Musical *Come from Away*, about residents of Gander, Newfoundland and Labrador, taking in stranded air travellers in the wake of 9/11, proved a hit on Broadway *(Steam Pipe Trunk Distribution Venue)*.

The True North: Images of Canadian Creativity

Clockwise from top: James Naismith, basketball's inventor; Toronto Raptors star Kawhi Leonard in Game 2 of the 2019 NBA finals (*Chensiyuan/Wikimedia Commons*); hundreds of thousands of fans pack Toronto's Nathan Phillips Square after Raptors' victory (*Andrew Scheer/Wikimedia Commons*).

The True North: Images of Canadian Creativity

Top right: Brian McKeever, winner of the most medals ever at the Paralympics *(Aldandra)*.

Middle right: Mikaël Kingsbury, most accomplished moguls skier of all time, at the 2015 Moguls World Cup *(Clément Bucco-Lechat)*. Bottom: Pioneering giraffe researcher Anne Innis Dagg, who is featured in the film *The Woman Who Loves Giraffes*, directed by Alison Reid and produced by Reid and Joanne Jackson *(Courtesy The Woman Who Loves Giraffes/Pursuing Giraffe Adventures Inc.)*.

The True North: Images of Canadian Creativity

Top left: Communications scholar Marshall McLuhan in 1945 *(Josephine Smith, LAC)*. Top right: Author Margaret Atwood in Stockholm, 2015 *(Frankie Fouganthan)*. Bottom left: Inuk throat singer Tanya Tagaq Gillis *(Joe Mabel)*. Bottom right: Acclaimed jurist Louise Arbour in 2011 *(World Economic Forum)*.

Chapter Five
The Arts and Entertainment

For many Canadians, creativity manifests itself more in the arts and entertainment than anywhere else. This is because composers, authors, playwrights, painters, singers, dancers, filmmakers, and others create the songs, stories, music, paintings, plays, films, dances, and other artistic and entertainment activities that say a great deal about Canada, Canadians, Canadian culture, and Canadian creativity.

While Canadians enjoyed many activities of this kind prior to the twentieth century, most of these activities were local, regional, amateur, and semi-professional rather than national, international, and professional in nature. However, during the second half of the twentieth century and first two decades of the present century, the arts and entertainment have blossomed considerably in Canada. In fact, many believe that Canada has been going through an artistic renaissance over the last 50 years that has been fuelled by the rapid expansion of these activities in all parts of the country, and, as a result, a phenomenal increase in the number of artists, arts organizations, activities, facilities, programs, and audiences as well as a great deal of international attention and acclaim.

This is most evident in popular music. Spurred on by the earlier successes of such well-known Canadian popular singers as Hank Snow (who is credited by some with helping Elvis Presley to start his career and inspiring Bob Dylan), Wilf Carter, Paul Anka, Andy Kim (Andrew Youakim), Giles Vigneault, Joni Mitchell, Leonard Cohen, Gordon Lightfoot, "Stompin' Tom"

Connors, The Band, Buffy Sainte-Marie, Bruce Cockburn, Anne Murray, Neil Young, The Guess Who, Rush, and Blue Rodeo, a new generation of popular singers has emerged in Canada over the last few decades that has taken the world by storm. Included here are such internationally acclaimed artists as Céline Dion, Shania Twain, k.d. lang, Bryan Adams, Corey Hart, Alanis Morissette, Sarah McLachlan, Avril Lavigne, Nelly Furtado, Susan Aglukark, Michael Bublé, Justin Bieber, Drake, The Weekend, Carly Rae Jepsen, The Barenaked Ladies, Nickelback, The Tragically Hip, and others. There is hardly a country in the world that hasn't been exposed to these and other Canadian popular singers showcasing Canada's huge talent pool in popular music.

While Canada's popular singers may not be credited with coining a uniquely "Canadian" sound or style—such as Elvis Presley did in the United States and the Beatles did in England—Canadian popular singers are well known for their international reach transcending the realm of music. Beyond their famous songs, many of these artists are recognized as influential world leaders for movements they have been involved in, changes that they have invoked, or breakthroughs that they have achieved, especially over the last few decades. In doing so, they have used their influence to help lead the charge on conversations about civil rights, indigenous rights, environmentalism, and gender issues.

Some obvious examples of this are such famous songs as "Both Sides Now" by Joni Mitchell and "Hallelujah" and "Suzanne" by Leonard Cohen; Anne Murray's influence in bringing country music into the mainstream; k.d. lang's multiplatinum 1992 hit "Constant Craving" and subsequent coming out, sparking a cultural revolution on perceptions of lesbianism; Neil Young's protest songs such as "Ohio" on the Kent State shootings and activism with the environment through organizations like Farm Aid; Buffy Sainte-Marie's numerous contributions to the promotion of Indigenous Rights and Indigenous music throughout the world; and Sarah McLachlan's founding in 1997 of Lilith Fair, national and international touring festivals exclusively featuring female solo singers and bands that did a great deal to advance the interests, visibility, and credibility of female singers and groups throughout the world. While many of these topics are front of mind today, Canadian musicians were pushing the boundaries and sparking a dialogue on these matters decades ago. Add to this the fact that four of the country's most popular male singers—Michael Bublé, Justin Bieber, Nickelback, and Drake—filled four of the five top spots of the first 200 spots on the American *Billboard* in 2011 and it is clear that Canada's international reputation in popular music is well deserved and much appreciated globally.

Quite recently, Drake has been extremely popular throughout the world. He was born Aubrey Drake Graham in Toronto in 1986. His father, Dennis Graham, was a black musician and well-known drummer with rock star Jerry Lee Lewis, and his mother, Sandra, was a white Ashkenazi Canadian Jew with a musical background as well, who raised Drake herself after his father went back to the United States. They were extremely poor when Drake was growing up and he was often bullied in school for being both black and Jewish. Ironically, it would be these Canadian multicultural influences that Drake would soon channel into his music to become a global pop star.

A major breakthrough occurred for this creative singer and songwriter when he was fourteen and got an important role in a popular Canadian television program called *Degrassi: The*

Next Generation about teens attending high school. It wasn't long after this that Drake dropped out of high school—although he eventually did finish high school and got his degree, which he said was "one of the greatest feelings in my entire life" — and began writing lyrics for songs as well as singing a great deal of rap music. One success followed another after this. The mixtape—*Best I Ever Had*—that featured Drake as the lead single quickly catapulted to number 2 on the Billboard Hot 100 chart and kicked off Drake's long reign over the charts. This mixtape remained eight straight years on the Hot 100—a run that eventually ended in 2017 with 431 consecutive weeks.

Drake's biggest break, however, was signing a major contract with Lil Wayne's label Young Money, which led in 2015 to the recording of his "Hotline Bling," a music video with over one billion views. This was followed, in 2016, with 13 music award nominations, which broke Michael Jackson's record of 11 in 1984, confirming Drake as the world's most popular recording artist, as well as his song "One Dance," which was Spotify's most-streamed song with over 882 million streams. In 2018, "God's Plan" broke Taylor Swift's record for most streams in a single day in the United States, and Drake topped the Spotify charts as the most streamed artist in that year globally with 8.2 billion streams for albums such as *Scorpion* and "God's Plan." Then, in 2019, Drake broke Taylor Swift's long-standing record for most wins at the Billboard Music Awards, with 12 wins and a career total of 27 compared to Swift's total of 23.

It is achievements like this that help to explain why Canada's reputation in popular music throughout the world is out of all proportion to the small size of the country's population. As Canadian broadcaster and music critic Alan Cross observed in talking about the success of Canada's popular singers,

"Canadians continue to punch above their weight."

While Drake symbolizes the success of Canada's popular singers internationally in some ways, David Foster symbolizes it in other ways. This multi-talented and highly creative artist from Vancouver has won numerous Grammy and Producer of the Year Awards, and is well known and respected throughout the world as a composer, arranger, song writer, producer, director, and mentor. He has written many popular songs that have been performed and recorded by some of the world's best known and most celebrated singers, including *All I Know of Love* (Barbra Streisand), *One More Chance* (Madonna), *The Colour of My Love, The Prayer,* and *To Love You More* (Céline Dion), *After Tonight* (Mariah Carey), *Because We Believe* (Andrea Bocelli), and others. Despite these remarkable achievements, Foster is much better known on the international music scene for the help he has provided over the years to many popular singers in different parts of the world when they have needed it the most, namely during the initial stage or stages of their careers and career development. Many of these singers have gone on to achieve international stardom and have become singing sensations in their own right.

What is true for Canada's popular singers is also true for its classical singers. Many of the country's most outstanding classical singers at present are recognized in musical circles around the world for their talents and accomplishments, such as Isabel Bayrakdarian, Measha Brueggergosman, Daniel Taylor, Russell Braun, Michael Schade, Ben Hepner, The Tenors, Karina Gauvin, Inuk throat singer Tanya Tagag Gillis, and others. The Tenors, for instance, achieved international acclaim for their appearances on *Oprah*, as well for winning numerous Emmy awards and other prizes. Much as in popular music, these achievements have not come out of the blue. They result from a solid tradition of earlier classical and operatic singers in Canada—a tradition that includes Jon Vickers, Lois Marshall, Léopold Simoneau, Louis Quilico, Maureen Forrester, Mary Morrison, Elizabeth Benson Guy, and Teresa Stratas. Many of these singers performed to large and enthusiastic audiences in the great opera houses of the world and provided motivation and inspiration to the more contemporary generation of talented Canadian singers and their accomplishments abroad.

In institutional terms, one of Canada's best known arts organizations internationally is the Canadian Opera Company. It has produced many outstanding operas in other parts of the world, and has capitalized on a rich reservoir of creative talents in Canada, such as Robert Lepage, the outstanding Quebec producer, director, and designer of theatre, opera, and music theatre who is known globally for his inventive staging of Bartok's *Bluebeard Castle* and Schoenberg's *Erwartung*—which won prestigious awards at the Edinborough Festival in 1993—as well as Stravinsky's *The Nightingale and Other Short Fables*, which premiered in Toronto in 2009 to great acclaim. This was a co-production with the Festival d'Aix-en Provence, Opera National de Lyon, and the Netherlands Opera in association with Ex Machina of Quebec. It was produced again in 2010 with the prestigious Brooklyn Academy of Music, one of the best-known music academies in the world, whose production of this work was deemed "pure magic" by the *New York Post*. Most recently, Lepage's superb staging of *Coriolanus* at the Stratford Festival received rave reviews from the *New York Times* in 2018.

Choral music and choral conductors should be added to this list of Canadian singers who have achieved international distinction and acclaim. Here, as well, Canada has achieved remarkable results. As with popular and classical singers, Canada has a well-established tradition of choral music that dates back more than a century, thanks to the keen interest in choral music in general and English choral music in particular, as well as the existence of countless church, community, and professional choirs across the country.

However, Canada was not really recognized internationally in choral music until Elmer Iseler arrived on the scene in the latter part of the twentieth century. According to Walter Pitman, author of *Elmer Iseler: Choral Visionary*, Iseler was a choral genius who not only created the world-renowned Festival Singers and Elmer Iseler Singers, but also rejuvenated the Toronto Mendelssohn Choir, one of the oldest choral institutions in Canada and the world that, along with the Toronto Choral Society, was established in 1845.

During his tenure, Iseler conducted numerous choral concerts in Canada, United States, Europe, and other parts of the world. His enormous presence in choral music has been sustained and augmented by many outstanding contemporary Canadian choral conductors who have followed in his footsteps, most notably Jean Ashworth Bartle, Robert Cooper, Linda Beaupré, and Lydia Adams in Toronto, Howard Dyck in Hamilton, Noel Edison in Elora, Gerry Fagan in London, and Jon Washburn in Vancouver. Thanks to Iseler's commitment and dedication to choral music and his vision and courage in featuring choral music by Canadian composers in many of his concerts, Canada now has a large and impressive repertoire of choral music, as well as many first-class ensembles that are making their mark on the global scene. Many of them are on a par with some of the finest choral organizations and choirs from Finland, Sweden, Russia, and elsewhere.

Canada is also well known throughout the world for the charismatic and eccentric pianist Glenn Gould who emerged in the middle part of the twentieth century and hummed while he performed in concert. Gould is recognized throughout the world for playing Bach and especially the *Goldberg Variations* brilliantly. He was and still is also well known for his audacity in telling Leonard Bernstein, conductor of the New York Philharmonic, how the Brahms piano concertos should really be played! His creative achievements on the piano—both in the concert hall and later in the recording studio because he gave up live performances in his all-too-short life—did a great deal to pave the way for a generation of remarkable Canadian pianists who are entertaining people all over the world and winning countless international awards, including Angela Hewitt, Marc-André Hamilin, Louie Lortie, André Laplante, Jane Coop, Janina Fialkowska, Anton Kuerti, Valerie Tryon, Stewart Goodyear, Jon Kimura Parker, Jan Lisiecki, and others.

Mention should also be made here of Tafelmusik and L'Opera Atelier. The former organization has travelled extensively throughout the world on numerous occasions, giving outstanding concerts and achieving international success for its performances of baroque and romantic music, often on original instruments and in churches and concert halls as well as on numerous recordings. The latter organization has followed a similar pattern and path, largely by staging highly successful operas by baroque and especially French baroque composers in the most recognized salons and venues of Paris, other parts of Europe, and North America in such places as the Royal Opera of Versailles, La Scala in Milan, the Salzburg Festival, and the Harris Theatre in Chicago.

These accomplishments were matched on the jazz side by the

incomparable Oscar Peterson, one of the world's greatest and most creative jazz musicians. Louis Armstrong called Peterson "the man with four hands" and Duke Ellington labelled him "the maharajah of the keyboard." These accolades were well deserved as Peterson had incredible talents on the keyboard. He became well known throughout the world soon after 1962 when he wrote and performed "The Hymn to Freedom." It quickly became a favourite of Martin Luther King Jr. and was embraced as the anthem of the American civil rights movement with its easy blues melody and remarkable improvisations on the piano by Peterson as well as its evocative words:

> When every heart joins every heart and together yearns
>   for liberty
> That's when we'll be free …

For this and for many other creative contributions, Peterson was accorded one of Canada's highest honours when a bronze statue of him by artist Ruth Abernethy was ensconced on Parliament Hill in Ottawa after his death.

Another accomplishment in music and related realms should be included here as it has had an important impact on the world in recent years. It is the contributions by R. Murray Schafer, one of Canada's most prominent composers, to the creation and development of the concept of "soundscape" and the new discipline of "acoustic ecology." According to Schafer, the world has become a "sonic sewer" over the last century as a result of the spread of noise pollution throughout the world due to new technologies, piped-in music, the colossal increase in vehicular traffic, and ever-higher decibel levels in neighbourhoods, communities, towns, and cities as well as in bars, restaurants, night clubs, festivals, and other venues and events. Schafer contends that this level of noise is very destructive to our hearing and our lives, with increased deafness and major hearing problems earlier and earlier in life.

It was largely due to this concern that Schafer created the idea of soundscapes in order to study the nature, frequency, and magnitude of sounds in different parts of the world in the hope that a better soundscape can be orchestrated and produced in the world of the future. This is documented in detail in his book *The Tuning of the World.*

Directly and indirectly, these activities led to the creation of the new disciple of acoustic ecology, as well as Schafer being recognized internationally as its founder. The purpose of this discipline is to place much more emphasis on reducing and hopefully eliminating obnoxious sounds and objectionable noises in all parts of the world, as well as creating and orchestrating a much more pleasant and enjoyable world soundscape in the years and decades ahead. One of Schafer's most valuable contributions as a composer is the creation of a great deal of "environmental music" and music theatre through his Patria series, which includes a number of major compositions performed outdoors or in the wild, such as *Music for Wilderness Lake, Princess of the Stars, The Enchanted Forest,* and others.

Remarkable achievements like this in music and related areas have been matched, if not surpassed, in literature. Canadian authors have been winning countless international awards and prizes in recent years. Here, as well, Canada's authors have been able to benefit from—and build on—a solid tradition in the literary arts stretching back to such popular writers as Thomas Chandler Haliburton, Susanna Moodie, Catherine Parr Traill, Louis-Honoré Fréchette, Frederick Philip

Grove, Louis Hérmon, Lucy Maud Montgomery, and others. This latter author, mentioned earlier, is cherished in Canada and around the world for such novels as *Anne of Green Gables* and *The Road to Avonlea* that centre on the life and experiences of a little orphan girl named Anne Shirley who grew up in Prince Edward Island under the supervision and guidance of foster parents.

Following on the heels of earlier generations of authors were generations of authors who appeared in the middle and latter decades of the twentieth century, including Morley Callaghan, W.O. Mitchell, Anne Hébert, Gabrielle Roy, and others. Morley Callaghan, whose concern for "the little guy" is legendary and is claimed to have defeated Ernest Hemingway in a boxing match, wrote about crime, punishment, deprivation, and a variety of secular and sacred matters in the country's rapidly evolving urban centres in books like *They Shall Inherit the Earth*, *Strange Fugitive*, and *The Loved and the Lost*. W.O. Mitchell, a much loved literary figure from western Canada who was also a great raconteur and humourist in the Stephen Leacock and Thomas Chandler Haliburton tradition, wrote about farm life on the prairies in books such as *Who Has Seen the Wind* and long-running CBC radio and television programs like *Jake and the Kid* and *Rawhide*. Anne Hébert, an outstanding Quebec poet, playwright, and novelist, wrote *Kamouraska*, which won the Prix des Libraries in France and was eventually made into a movie by the well-known Quebec filmmaker Claude Jutra. Gabrielle Roy, another well-known French-Canadian author, but this time from Manitoba rather than Quebec, wrote *Bonheur d'occasion* (*The Tin Flute*) (1945) and *La Petite Poule d'eau* (*The Little Water Hen*) (1950), both of which won a number of coveted literary awards in France.

These authors were followed by Antonine Maillet, who wrote about the expulsion of the Acadians from Nova Scotia in 1755 and Acadian life in Canada and elsewhere in the world in *La Sagouine* and other books; Margaret Laurence, whose two classic novels *The Diviners* and *The Stone Angel* were made into movies; and Mordecai Richler, who is well known for books about growing up in the side streets and back alleys of Montreal such as *The Apprenticeship of Duddy Kravitz* and *St. Urbain's Horseman*.

While traditions such as these help to explain why contemporary Canadian authors have been winning many coveted international awards and prizes in recent years, the most important factor of all is the quality, creativity, and originality of the authors themselves and their literary works. Take Margaret Atwood, for instance, one of Canada's most prolific, internationally known, and respected authors, who is appreciated throughout the world for her many books and novels on a variety of feminist themes, mythological matters, and human and social problems, such as *Survival, Surfacing, The Edible Woman, The Handmaid's Tale, Cat's Eye, Alias Grace, The Blind Assassin, Oryx and Crake*, and others. Many of these books have won prestigious national and international literary awards and been made into movies or television series, such as *Alias Grace* and *The Handmaid's Tale* documented in detail earlier.

Other Canadian writers have also received a great deal of international attention and recognition. Michael Ondaatje, a Sri Lankan of Dutch-Indian descent who emigrated to Canada in the early 1980s, is the author of many books, including *In the Skin of a Lion* and *The English Patient*, which won the Man Booker Prize in 1992 and was made into a movie. Yann Martel won the Man Booker Prize in 2002 for his book *The Life of Pi,* which was also made into a movie; Alistair MacLeod won the IMPAC Dublin Literary Award for his novel *No Great Mischief* in 2001; Carol Shields won ten Pulitzer, Orange, and Governor-General's awards for such books as *The Stone Diaries, The Orange Fish,* and *Dressing Up for the Carnival*; and most recently, Esi Edugyan, who won her second Giller Prize for *Washington Black* in 2018.

It is books and movies like these that are making it possible for people in other parts of the world to become familiar with important aspects of Canadian history, geography, traditions, culture, stories, and literature. Interestingly, Canada is also becoming well known throughout the world for the ethnic and racial diversity of its authors. In addition to those already mentioned, there are also such notable authors as Joy Kogawa, M. G. Vassanji, Austin Clarke, Madeleine Thien, Lawrence Hill, George Elliott Clarke, and Marlene Nourbese Philip, to name but a few among many.

The crowning achievement in this area was undoubtedly the awarding of the Nobel Prize for Literature to Alice Munro in 2013. In many ways, this award and author epitomize the rich creativity that Canadian writers have manifested over the last few decades, which is why so many of these writers were quick to point out that Alice Munro's literary accomplishments, which also included winning the 2009 Man Booker International Prize and three Governor General's awards, were "monumental" and in a "class by themselves." Munro is internationally known as someone who has revolutionized the short story—"Canada's Chekhov" is how one author described her—through writing about life and living in Huron County in southwestern Ontario and other small towns and rural areas in Canada in books such as *Too Much Happiness*, *The Love of a Good Woman*, *Dance of the Happy Shades*, and others.

Canada is also internationally known for its children's literature. The country's authors and publishers have been in the vanguard of many global developments in this field. Notable organizations and publishers include Kids Can Press, CitizenKid, the Canadian Children's Book Centre, Second Story Press, and Tundra Books. CitizenKid, for instance, is a collection of branded books published by Kids Can Press (part of the Canadian media company Corus) that is predicated on making children more aware of developments throughout the world and helping them to become global citizens. The collection sold rapidly in 24 countries around the world including Japan, Turkey, and Korea, and was translated into more than 20 languages. In recent years, these developments have been accompanied by many creative achievements in television for children through Kids' TV, Treehouse, YTV, Corus, and others.

Canadians have been also actively engaged in the creation of comic books from the very outset. As far back as the 1930s, Joe Shuster, who was born in Toronto and grew up in this city before moving to the United States, created the comic book hero Superman in collaboration with an American, Jerry Siegel. Superman first appeared in issue number one of *Action Comics* in 1938, and has since been featured in newspaper strips, TV shows, movies, and of course comic books or, as they are more often called today, graphic novels. In recognition of this and other contributions to the development of the comic book industry, the Canadian Comic Book Creator Awards Association created the Joe Shuster Awards in 2005 to honour this Canadian-born pioneer.

But the contribution of Canadians to the creative

development of comic books and the comic book industry does not end here. Far from it. Canadians Harold Foster, Dave Sim, and Tod McFarlane were actively engaged in the creation of internationally popular characters in this field such as Prince Valiant, Cerebus the Aardvark, and Spawn. Cartoonist and author Charles Thorson from Winnipeg worked at several American animation studios including the Walt Disney Company and Warner Brothers where he credited with creating Elmer Fudd, Snow White, and many other characters. Adrian Dingle created Nelvana of the Northern Lights; Stanley Berneche introduced the world to *Captain Canada*; Leo Bachle gave us *Johnny Canuck*; and there were many others of note, such as Chester Brown, Joe Matt, Seth, Dave Collier, and Julie Doucet. John Bell's informative book *Invaders from the North: How Canada Conquered the Comic Book Universe,* published by Dundurn Press in Toronto in 2006, profiles past and present comic book geniuses and sheds a great deal of light on a much-neglected chapter in pop culture history. Bell demonstrates how Canadian creative talents vaulted Canada into the forefront of the international comic book industry in both artistic and commercial terms, and, in so doing, challenged many long-standing traditions and time-tested boundaries between high and low culture in Canada and other parts of the world.

It may come as a surprise to many, but Canada and Canadians have also made numerous creative contributions to the development of the film industry over the last 125 years that have had a major impact on the world. While it was a Frenchman, Monsieur Lumière from Lyon, who brought the first moving picture to Canada in 1896, two Canadian brothers—Andrew and George Holland from Ottawa—were the first to open a movie house in North America. It was called a "kinetoscope parlour," and opened in the United States in 1894, grossing over $16,000 in its first year. About the same time, James Freer purchased a combination camera, projector, and printer and became Canada's (and one of the world's) first filmmakers after viewing the first public screening of motion pictures by Auguste and Louis Lumière in the basement of a café on the Boulevard des Capuchines in Paris in 1895. Freer returned to Canada and made a number of short films about trains and farm life in Manitoba where he and his family lived in the late nineteenth century. In addition, Ernest Ouimet, a Canadian blue-collar worker who eventually became an entertainment mogul, opened the Ouimetoscope Theatre in Montreal shortly after this. It was followed in 1907 by a second Ouimetoscope Theatre, which had over a thousand upholstered seats, a refreshment bar, and an orchestra. It was the first large-scale moving-picture palace—called a "grind house" at the time—in North America.

By the first two decades of the twentieth century, many Canadians were heading off to the United States to get involved in the creation and early development of Hollywood. One of the most important of these was Sidney Olcott, who was born in Toronto. As an actor and screenwriter, but more importantly as a filmmaker, producer, and director, Olcott was active from the very start of Hollywood, making several silent films, including the first version of *Ben Hur* in 1907 and *From the Manger to the Cross* in 1912. The latter film grossed over a million dollars and is claimed by some to have had a major impact on the development of such world-famous film directors as D. W. Griffith and Cecile B. DeMille.

Olcott was viewed by many as the greatest film director of this era. He paved the way for many other Canadians, including those who established world-famous movie companies, most notably Louis B. Mayer from New Brunswick who founded Metro-Goldwyn-Mayer Studios and came up with the idea of the Academy of Motion Picture Arts and Sciences, and Jack Warner from London, Ontario who founded Warner Brothers Studios. The latter employed some of Hollywood's greatest actors and actresses, including Betty Davis, Errol Flynn, James Cagney, Edward G. Robinson, and Humphrey Bogart. To this list should be added Mack Sennett from Danville, Quebec who founded the Keystone Kops and was instrumental in establishing the careers of comedians Charlie Chaplin and W.C. Fields, as well as paving the way for the comedy films made by Bing Crosby and Bob Hope. These distinguished directors, and other Canadians, also played an important role in the making of such classic films as *Casablanca, Gone With the Wind, The Wizard of Oz,* and many of the Charlie Chaplin, Buster Keaton, Gloria Swanson, and Marie Dressler movies. Their contributions, and those of others, are described in detail by another fascinating individual who had a long and strong association with Canada and Hollywood, namely Charles Basil Foster, who was born in England but spent most of his life in Canada. Primarily a publicist, Foster worked closely with such well-known actors and actresses as Boris Karloff, Marilyn Monroe, and Richard Burton. Two of his books document in detail the remarkable impact Canadians had on the origins and early development of Hollywood: *Stardust and Shadows*, which is about the silent film era; and *Once Upon a Time in Paradise*, which is about Canada's contributions to Hollywood from the late 1920s to the 1950s.

This by no means completes the story of Canada's creative contributions to the development of Hollywood, as well as to the film industry in general. When the National Film Board (NFB) was created in 1939 to provide Canadians and people in other parts of the world with information about Canada and its involvement in the Second World War, few would have thought that it would soon become one of the largest and most prestigious public film agencies in the world. Through its mandate to "produce, distribute and promote the production and distribution of films designed to interpret Canada to Canadians and to other nations," the NFB made many highly original contributions to the war effort as well as to the art of filmmaking. This includes creative contributions to the making of film documentaries, cinéma vérité (literally "truth in film" or "truth films"), split-screen and multi-screen viewing, the training of thousands of male and female filmmakers, directors, and technicians, and especially film animation.

Many of the NFB's early contributions to film animation were made by Norman McLaren. He was without doubt one of the greatest pioneers in the early development of animation and won many international awards for his efforts, including Academy awards in the United States and Britain, the Palme d'Or at Cannes, and a number of first prizes at film festivals in Venice as well as the Silver Bear in Berlin. In addition to this, who would have thought that millions of people around the world would watch NFB films every year for many years after the NFB was created? By the end of World War II, the NFB was the largest documentary filmmaker in the world, with some 800 employees in strategically located offices in London, Washington, New York, Chicago, Mexico City, and Sydney and more than 500 films to its credit.

Over the last century, Canada has produced many very talented and highly creative male and female documentary filmmakers, not only at the National Film Board—where many male and female filmmakers got their start—but also throughout the private and public sectors. It all started with the production of *Nanook of the North* in 1922 and popular NFB film documentaries such as *Canada Carries On*, *Lest We Forget*, and *World in Action*, which was shown in 6,500 cinemas throughout the world during and after the Second World War. This was followed by McLaren's documentary *Neighbours*, which won an Academy of Motion Picture Arts and Sciences award, as well as the highly innovative film documentary *Challenge for Change* about the plight of fishermen in Newfoundland, many of whom were involved in making this documentary themselves and playing important roles in it.

After the Second World War, documentary filmmaking blossomed in all parts of Canada and made a strong contribution to the making of documentary films internationally by such talented filmmakers as Allan King, F. R. "Budge" Crawley, Donald Brittain, Don Owen, Robin Spry, Harry Rasky, Bonnie Sherr Klein, Alanis Obomsawin, Holly Dale, Rombus Media, Anne-Claire-Poirier, and many others. Most recently, Canadian documentary filmmakers Jennifer Baichwal, Michael de Pencier, and Ed Burtynsky created and produced the documentary film *Anthropocene: Human Epoch*, which has won numerous awards as a documentary about the incredible impact humans have had and are having on the natural environment and the world generally. This and other talented Canadian filmmakers have confirmed John Grierson's belief that Canadians are capable of making a major contribution to the world through documentary filmmaking, animation, cinéma verité, and other activities.

And this is not all. Since the Second World War, many of Hollywood's most outstanding and revered directors, producers, and screenwriters have been or are Canadian, including Arthur

Hiller, Allan King, Alan Thicke, Bernard Slade, David Steinberg, Harry Rasky, Ivan Reitman, Lorne Michaels, Ted Kotcheff, Norman Campbell, James Cameron, Norman Jewison, and others. Also noteworthy are film composers such as Howard Shore, who has won three Academy Awards and wrote the music for the *Lord of the Rings* trilogy and many other well-known films.

Many of Hollywood's best-known actors and actresses since the early years have also been, or are, Canadian. This includes such international stars as Lorne Greene, Pa in *Bonanza*, William Shatner, Captain Kirk in *Star Trek*, and Christopher Plummer, co-star with Julie Andrews in *The Sound of Music*. And this is only the tip of a much larger iceberg. A more complete list of Canadian actors and actresses who have made it big in film and Hollywood over the years includes Mary Pickford, Marie Dressler, Walter Pidgeon, John Ireland, Walter Huston, Raymond Massey, Raymond Burr, Fay Wray, Norma Shearer, Glenn Ford, Barry Morse, Chief Dan George, Jay Silverheels, and Robert Goulet in an historical sense, and Donald Sutherland, Kiefer Sutherland, Michael J. Fox, Dan Aykroyd, Geneviève Bujold, Kim Cattrall, Victor Garber, Mike Myers, Jill Hennessy, and many others in a contemporary sense.

Mention should also be made here of IMAX, which was created by Canadians as a motion picture device that possesses the capacity to record and display images of far greater size and resolution and on much larger screens than conventional film display systems. This innovative technology was developed by Canada's IMAX Corporation, and especially by Graeme Ferguson, Roman Kroitor, Robert Kerr, Nicholas Mulders, and William C. Shaw. IMAX was first displayed at Expo '70 in Osaka, Japan, and had its first permanent display at Ontario Place in

Toronto in 1971. It is now a standard feature of the film industry in all parts of the world.

Capitalizing on earlier successes and opportunities provided by previous generations of Canadians and especially by the National Film Board, many outstanding filmmakers and directors from Quebec including Claude Jutra, Michel Brault, Denis Héroux, and Denys Arcand have achieved international prominence for their creativity in filmmaking. Arcand has been especially successful in this respect, winning a number of Quebec and international awards in recent years for such films as *Jésus de Montréal* (*Jesus of Montreal*), *Le Déclin de l'émpire américan* (*The Decline of the American Empire*), *Les invasions barbares* (*The Barbarian Invasions*), which won an Academy Award for Best Foreign Language Picture in 2003, and, very recently, *The Fall of the American Empire* in 2018. During the last few decades, many other outstanding Quebec filmmakers have emerged such as Xavier Dolan, Denis Villeneuve, Jean-Marc Vallée, and Philippe Falardeau who are enjoying a great deal of success on the Canadian and international film scene.

Unlike Quebec filmmakers and directors who have been able to capitalize on producing films in French for largely French-speaking audiences, filmmakers and directors in other parts of Canada have found it far more difficult to make, finance, and promote their films. This is because their films are produced in English and must compete with American films, as well as the fact that most movie houses in Canada are owned and operated by huge American entertainment and movie conglomerates that have little interest in Canadian films and filmmakers. Nevertheless, some well-known and highly creative filmmakers and directors in other parts of the country have achieved significant results in international terms over the last three or four decades, including David Cronenberg, producer of *The Fly*, *Naked Lunch*, and the controversial film *Crash*, Atom Egoyan, a talented film director of Armenian descent who produced *Erotica* and the equally controversial *Ararat*, Deepa Mehta, a film-maker of East Indian origin who has produced many popular films such as *Water* that won several awards, Zacharas Kunuk, an Inuit whose *Ajanarjuat: The Fast Runner* received a great deal of critical acclaim at film festivals around the world, and Sarah Polley, a well-known actress in Canada for her role in the *Road to Avonlea*, who has directed a number of popular television series and films recently, especially *Away from Her* about a woman suffering with dementia.

Mention should also be made here of the Toronto International Film Festival and the impact it has had on the world since it was created in the 1970s. Generally regarded as the second most important film festival in the world after the Cannes Film Festival in France, it attracts thousands of actors, actresses, directors, designers, producers, entrepreneurs, buyers, and sellers from all over the world who are anxious to see the latest and many of the best films that have been produced in the world. According to Roger Martin, dean of the Rotman School of Management at the University of Toronto, TIFF, as it is affectionately referred to by many Canadians, is one of the top globally significant Canadian brands, along with the Four Seasons Hotels, the Cirque du Soleil, and the BlackBerry.

Unfortunately, one of Canada's most creative achievements in the arts is not well known in other parts of the world (although it is high time it should be). It is the artistic works of the Group of Seven. Following in the footsteps of Tom Thomson, who is generally regarded as Canada's greatest landscape painter, the Group of Seven was formed in the 1930s and '40s and became well known throughout Canada for rejecting formal styles and traditions in painting and creating a style that was more "distinctively Canadian." While they incorporated some elements of Impressionism, which was popular at the time, they ended up painting in a style that is very recognizable and quite distinct.

The Group travelled to many parts of Canada and into the far north to paint the country's majestic lakes, rivers, mountains, forests, landscapes, seascapes, tundra, and wilderness areas. Comprised of Franklin Carmichael, Lawren Harris, A. Y. Jackson, Frank H. Johnston, Arthur Lismer, J. E. H. MacDonald, and Frederick Horsman Varley, the Group had a vision of Canada—and a keen understanding of the role that artists and the arts should play in the development and promotion of that vision—that they were anxious to share with other Canadians and the world at large. While Emily Carr was not a member of the Group and her style of painting was different from that of the Group, she had a close affiliation with several of its members and her aesthetic talents were instantly recognized and embraced by the Group, especially her depictions of nature, landscapes, and totem poles carved by the Indigenous peoples of British Columbia.

What was most significant about the Group of Seven at the time was the effect it had on the development of the arts in Canada and the country as a whole. The Group's belief that the arts must flourish before Canada will become a real home to Canadians struck a responsive chord with people in the cultural community and indeed many other Canadians. Slowly but surely, Canadian artists and increasing numbers of people in other fields began to see Canada and its various problems and possibilities in a new light—a light that focused much more attention on Canadian customs, traditions, and values and less on European and American customs, traditions, and values. As such, the Group set the stage for the escalating interest in the arts and Canadian artists—and especially in "made in Canada" rather than "imported" developments—which is evident in more countries throughout the world today.

Despite the lack of attention historically accorded to the Group of Seven in other parts of the world, things are starting to change—and change significantly—due to the quality, abundance, and relevance of the Group's artistic works and its basic philosophy, particularly with respect to nature, landscapes, the natural environment, and the role of the arts in society. Ian Desjardin, director of the Dulwich Picture Gallery in London, England—Britain's oldest public art museum—recently mounted a comprehensive exhibition of the works of the Group of Seven called *Painting Canada: Tom Thomson and the Group of Seven*. It was the largest exhibition of works by Tom Thomson and the Group of Seven ever mounted in the world with over a hundred paintings in the exhibition. As Desjardin claimed, "They're just beautiful—*so* beautiful. How can anyone not fall in love with them?"

Despite the lack of knowledge and awareness of Tom Thomson and the Group of Seven in other parts of the world, Canada is known internationally today for a number of its other artists, especially artists from Quebec who went to live and work in Paris after the Second World War. The impetus for this was provided by Alfred Pellan and a group of painters known as the *Automatistes*, several of whom were engaged in writing *Refus global*, which contributed so much to the transformation that took place in Quebec after the Second World War when the arts and culture were liberated from religion, politics, and the Church. This group included Jean-Paul Riopelle, Marcel Barbeau, Fernand Leduc, and Paul-Émile Borduas. They became well-known for abstract expressionism and geometric patterns and designs, as well as for paving the way for the next generation of painters in Quebec that included Guido Molinari, Claude Tousignant, Yves Gaucher, and many others.

While these developments were taking place in Quebec, a new generation of Canadian visual artists was coming to the fore in other parts of the country, including David Milne, Jack Bush, Michael Snow, Harold Towne, Jack Shadbolt, Roy Kiyooka, Kazuo Nakamura, Alex Colville, and, more recently, Mary and Christopher Pratt, Jeff Wall, Philip Guston, Douglas Coupland, Doris McCarthy, and many others. While it seems to take much longer for visual artists to make their mark on the world, there is no doubt that the works of these and other contemporary Canadian artists are becoming much better known in the world, something that augers well for the future. Canada is also becoming better known in other parts of the world for its wildlife artists, especially Glen Loates, Freeman Patterson, and Robert Bateman and their many paintings and prints. This is also true for many Indigenous painters, carvers, and printmakers, most notably Pitseolak Ashoona, Karoo Ashevak, David Ruben Piqtoukun, Joe Talirunili, Norval Morrisseau, Benjamin Chee Chee, Allen Sapp, Bill Reid, and others, as well as places such as Cape Dorset and Nunavut that are not only well known but also revered around the world for the quantity, quality, and excellence of their artistic endeavours, aesthetic sensibilities, and carvings, many of which are in great demand throughout the world today.

Humour and comedy are also areas where Canada has become well-known internationally and is making its mark.

In fact, many people around the world are probably as aware of Canada and Canadians for their humour and comedy as for their hockey.

It all started almost two centuries ago when Thomas Chandler Haliburton wrote Canada's first internationally known and highly successful book—*The Clockmaker*—which expounded on the exploits and escapades of the notoriously funny Samuel Slick. It was published in 1836 and was Canada's first foray into the realm of humour but certainly not its last. Later well-known humourists include Stephen Leacock and his hilarious *Sunshine Sketches of a Little Town*, Johnny Wayne and Frank Shuster and their popular performances on the *Ed Sullivan Show* in the United States as well as here at home on the CBC, Al Waxman and his situation comedy *King of Kensington*, and Mordecai Richler and his books *The Apprenticeship of Duddy Kravitz* (mentioned earlier) and *Jacob Two-Two Meets the Hooded Fang*.

With historical precedents like these, it is no coincidence that Canadian humour and comedy have blossomed substantially in recent decades. Countless Canadian comedians have achieved international success and fame for their creative accomplishments, including Mike Myers, Leslie Nielsen, Dan Aykroyd, John Candy, Jim Carrey, Michael J. Fox, Martin Short, Tom Green, Rick Green, Yvon Deschamps, Colin Mochrie, André-Philippe Gagnon, Russell Peters, Luba Goy, Ian and Will Ferguson, Ron James, Rick Mercer, and Dashan (Big Mountain), who is exceedingly popular in China as a comedian despite the fact that few Canadians know anything about him. Various comedy troupes and TV shows have also garnered international recognition, particularly *CODCO, Second City TV* (*SCTV*), the Royal Canadian Air Farce, *This Hour Has 22*

Minutes, the *Trailer Park Boys*, the *Red Green Show*, *Corner Gas*, Yuk Yuks, the *Rich Mercer Show*, *Little Mosque on the Prairies*, and, most recently, *Kim's Convenience* and many others. In fact, it may well be that Canada has become the "comedy capital of the world" with more well-known comedians than any other country.

If Canadians have made many creative contributions to comedy and humour that are known internationally, this is equally true for the circus and certain types of theatrical events. It would not be far off the mark to say that Canadians have virtually reinvented the circus in recent years, due largely to the efforts of the world-famous Cirque du Soleil.

This highly creative organization was founded in Quebec in the early 1990s by Guy Laliberté, Daniel Gauthier, Gilles Ste-Croix, Rachel Vertus, and others. Since then, it has taken the world by storm, primarily through its ability to use extraordinarily inventive choreographic techniques and a great deal of artistic wizardry and aesthetic mastery to transform the circus from the traditional spectacle involving clowns, elephants, and lion tamers into the most brilliantly conceived and exquisitely executed performances of artists and acrobats in a rich variety of acts, venues, and sites. Renowned for the performances' surreal atmospheres, awe-inspiring stunts, and incredible feats of bravery, daring, brilliance, physical endurance, and mental toughness, the Cirque has astounded and moved audiences in all parts of the world.

Not only is the Cirque now Canada's major entertainment company, but also it is the largest theatrical producer in the world, with over 4,000 employees from 40 countries, over 90 million people having seen its shows worldwide, and a budget of well over US$ 800 million. Its plans in the near future include the construction and operation of a major theatre in Hangzhou, China, the creation of a Cirque du Soleil Theme Park in Mexico, a major venue and theatrical production in Dubai, and, most recently, the acquisition of a major troupe of magicians called the Illusionists which includes *America's Got Talent* champion Shin Lim. And this is not all. Through the creation of a separate company, 45 Degrees, as a division of Cirque du Soleil, it also creates exclusive content, immersive experiences, and customized events and performances for a broad range of private companies and public institutions. One of its most popular attractions is *Saltimbanco*, which literally means "jump on the bench" in Italian. It is one of the longest-running touring shows of the Cirque, and has travelled to South America, Australia, Japan, South Africa, and many other destinations in the world.

Given the dynamic development and vitality of the arts and media in Canada, it is not surprising that the country has also made a number of seminal contributions to the administration of the arts and media as well as the training of arts and media managers, administrators, and policy-makers that have had an important impact on the world.

Canada's greatest contribution in this area has been the creation of a unique political and governmental system for public administration and funding of the arts and "cultural industries" that is seen as a model in other parts of the world. Ironically, it has evolved largely as a result of the inability of Canadians to decide whether the French or the British system of public administration and funding of the arts and cultural industries provides the most effective model for Canada and Canadians.

It is a well-known fact in artistic and media circles around the world that the French system of public administration and funding involves the creation of ministries of culture in government, whereas the British system involves the creation of arts councils situated at arm's length from government and the political process. For some curious reason, the French have never been too concerned about governments getting directly involved in a variety of artistic and media matters through ministries of culture. However, the British have always been wary of this scenario because such an arrangement in their view possesses the potential for interference with artistic freedom and the ability of artists and arts and media organizations to create activities, products, and programs of their own choosing.

Given the existence of large French and English populations in Canada, a hybrid model has emerged throughout the country over the last five or six decades with the federal government and many provincial governments having *both* a ministry of culture in most cases today, as well as an arm's length arts council. At the national level, for example, there is the Department of Canadian Heritage—which operates very much like a ministry of culture in everything except name—and the Canada Council for the Arts, which operates at some distance from the federal government. This pattern is repeated in most provinces as well as in many municipalities.

While this system has its problems—there are the obvious ones such as some duplication of activities and funding, sorting out who is responsible for what, and the potential for friction over duties, responsibilities, and jurisdictions—it also has numerous advantages. There is direct representation of the interests and concerns of artists and arts and media organizations in government and the political process, as well as arm's length treatment of artists and organizations in matters related to artistic freedom, excellence, funding, and independence from governments in general as well as from governmental regulations and political control in particular. It is a system that has worked well on the whole in Canada, and is also being examined by other governments and countries around the world.

Interestingly, Canada was also one of the first countries in the world—if not *the* first country—to provide substantial public funding for artists as individuals, largely through the Canada Council for the Arts, which was created in 1958 to provide public support for artists, arts organizations, and the development of the arts and media in Canada. Through its comprehensive system of senior, junior, and special grants to artists as individuals, the council sent out a clear signal to the rest of the world that Canada was serious about its intention to be creative when it came to public administration and funding of the arts, artists, arts organizations, and the media or cultural industries.. This intention has been followed since that time by other arts councils in Canada as well as by funding and administration departments and agencies in other parts of the world.

Canada's creative contributions to arts and media administration throughout the world have also been advanced considerably over the last four or five decades by the creation of academic training programs in this area at universities and community colleges across the country. This began in 1969 when a Program in Arts Administration was created at York University in Toronto. This program was the first comprehensive academic program for training arts managers, administrators, and policymakers in the world. It was established at the graduate level in the Faculty of Administrative Studies in 1970 and began providing courses for students in 1971. It is now called the Program in Arts, Media, and Entertainment Management as it is up-to-date with developments in this rapidly changing and extremely dynamic field, and celebrated its fiftieth anniversary in 2019 as a fundamental part of York's Schulich School of Business. The program combines practical training on the job with academic courses at the university. It also produces specialized reports and convenes seminars and workshops on administration, funding, and policy developments in the arts, media, and entertainment that are relevant to people in Canada and other parts of world.

What is most interesting about Canada's leading role in this area is the number of academic programs for training arts and media administrators that now exist throughout the world. It is estimated that there are more than 400 academic programs of this general type in various languages at universities, community colleges, and other educational institutions throughout the world today, whereas there were only two or three academic programs like this in existence in the early 1970s. This reflects the rapid growth of the arts internationally during this period, as well as the dire need for well-trained and highly qualified arts and media managers, administrators, and policymakers in all parts of the world.

The many creative contributions Canada and Canadians have made to the world in the arts and entertainment constitute a remarkable achievement when examined collectively, especially in view of the small size of the Canadian population compared to the populations of many other countries. It is a contribution that is likely to grow substantially in the years and decades ahead.

Chapter Six
Activism and Advocacy

Activism and advocacy are essential elements in the development of every country and culture in the world. Despite the problems caused by activists and advocates, their activities often bring about badly needed change and fundamental improvements in many if not all the diverse countries and cultures of the world.

Think of the remarkable impact that activists and advocates such as Karl Marx, Mahatma Gandhi, Dr. Martin Luther King Jr., Nelson Mandela, Ralph Nader, Mother Teresa, Paulo Freire, Dr. Sun Yat-Sen, and numerous others have had on specific countries and cultures as well as the world as a whole. Their efforts have changed many things in the world, including major advances in the lives of working class people, bringing home rule to India, recognizing the rights of blacks in America, making South Africa a much more inclusive country, providing consumers with safer products, challenging people to "give until it hurts," bringing about educational reforms in Brazil and other parts of the world, and instituting the overthrow of the Qing dynasty and paving the way for the creation of China as an independent country.

The techniques, tools, and means these and other activists and advocates used to achieve their aims and objectives are many and varied, and have been both peaceful and pleasant as well as provocative and painful. They run the gamut of possibilities, from writing letters to newspaper editors, organizing strikes, lock-outs, sit-ins, and walk-outs, and undertaking protests and marches to committing acts of civil disobedience, burning people in effigy, and doing what is often called "culture-jamming," or deliberately refusing to go along with the traditional way of doing things and centuries-old customs through carefully crafted forms of resistance, non-cooperation, and even violence if necessary.

For a country with a small population, Canada has had more than its fair share of activists and advocates. Fortunately, most of their activities have been peaceful rather than painful. While the focus in this chapter is on Canadians whose activism or advocacy has had an important impact on other countries and the world as a whole, they have all been involved in the development of creative ideas, activities, and projects as well as the quest to translate these ideas, activities, and projects into realities in one form or another. In the process of doing this, they have been engaged in many different causes and participated in the creation of many new programs, projects, and organizations that have had an important impact on the world or significant parts of it.

Of all the areas where Canadian activists and advocates have realized this, none is more significant or conspicuous than the contributions they have made to helping people with disabilities of many different types. And few if any people had a greater impact in this area in a global sense than did Jean Vanier, son of Major-General Georges Vanier (the 19th Governor General of Canada) and his wife Pauline Vanier. While some people might call Jean Vanier a social worker or humanitarian rather than an activist or advocate because of the approach that he took, there is no doubt that he played a major role in helping people with developmental disabilities throughout many parts of the world, and that he did so through creativity, commitment, and perseverance.

Jean Vanier was born in Geneva, Switzerland in 1928 when his father was carrying out a diplomatic mission there. In his youth, he served in the Royal Navy and the Royal Canadian Navy for a time before studying philosophy and theology at the Monastery of Le Saulchoir and the Institut Catholique in France in addition to spending a great deal of time meditating and thinking about his future.

While teaching at a number of universities in Europe following his academic studies, Vanier became interested in the plight of the millions of people who were suffering with disabilities in a variety of institutions throughout the world. This led him to invite two such people—Raphael Simi and Philippe Seux—to leave the institutions they were living in at the time and come and live with him at his home in Trosly-Breuil north of Paris. Vanier called his home L'Arche—The Ark. It became so popular that in 1964 he decided to create an organization called L'Arche in conjunction with Simi and Seux that would be devoted to the creation and growth of homes, programs, and support networks for people with developmental disabilities, particularly intellectual disabilities.

Unlike organizations that operate on a profit basis and are "client based," L'Arche operates on a non-profit basis and is "community based." People with disabilities and those who assist them live in homes and apartments provided in these communities, share their lives, and build relationships with people like themselves. As a faith-based organization whose motto is "changing the world, one heart at a time," L'Arche is predicated on the following principles and convictions as set out in the Charter of the Communities of L'Arche:

- Whatever their strengths or their limitations, people are all bound together in a common humanity;

- Everyone has the same dignity and the same rights, including the right to life, to a home, to work, to friendship, and to a spiritual life;
- A truly just and compassionate society is one which welcomes its most vulnerable citizens, and which provides them with opportunities to contribute meaningfully to the communities in which they live;
- Systems of belief—be they secular or religious—make the world a better place only when they promote the dignity of all human beings, inspiring us to be open to people of different intellectual capacities, social origins, races, religions, and cultures.

As a result of these principles, L'Arche communities endeavour to ensure that every person in the world has the ability to grow and mature into adulthood and make a valuable contribution to society regardless of their difficulties or limitations, primarily by developing long-term, mutually interdependent relationships, maintaining a stable, loving, and caring home life and friendly environment, and cooperating with care givers and other professional providers. As stated in their Charter, "In a divided world, L'Arche wants to be a sign of hope. Its communities, founded on covenant relationships between people ... seek to be signs of unity, faithfulness, and reconciliation."

L'Arche became so popular after it was founded that it has grown from its modest beginnings in France into a world-wide organization called the International Federation of L'Arche. The Federation has more than 5,000 members, with some 154 communities and official projects operating on every continent and in more than 38 different countries. The first community to be established in Canada was in Richmond Hill. It was created in

1969 and since that time has mushroomed into 29 communities across the country from Cape Breton in the east to Vancouver Island in the west.

In 1968, following a retreat held at Mary Lake outside Toronto, Vanier and his supporters created a companion international organization called Faith and Light. This organization provides retreats and prayer time for children with developmental disabilities, their parents, and adults who suffer from disabilities. It had a rocky beginning, since most hotels and motels would not provide accommodations to children and adults with developmental disabilities for fear that they would cause too much damage to their rooms and furniture. So Vanier and a friend named Marie-Hélène Mathieu decided to organize a huge pilgrimage for children, adolescents, and adults with developmental disabilities and their families and friends to Lourdes, France, the site of the Miracle of Lourdes where the Virgin Mary is claimed to have visited three girls.

After several years of preparation, Lourdes played host to more than 12,000 pilgrims from 15 countries in 1971, among them 4,000 people with developmental disabilities. Like L'Arche, Faith and Light has grown rapidly throughout the world since that time. It presently numbers more than 1,500 communities organized in 53 provinces located on five continents and operates in 83 countries and 38 different languages.

In addition to his numerous responsibilities with L'Arche and Faith and Light, Vanier found time to write many books that had a considerable impact in many different parts of the world. He espoused the belief that every person has important contributions to make to society.

Vanier attracted a large international following prior to his death at age 90 in 2019. Thanks to his efforts and those of other activists and advocates in this area, it is now more common to integrate people with developmental disabilities into society and to make them participants in the neighbourhoods, communities, and organizations where they live, work, and study.

Vanier's reputation and following made the revelations included in an independent consultant's report commissioned by L'Arche and released in 2020 all the more troubling. The report revealed allegations by six women (five of whom were directly interviewed by the consultant) that they had been sexually abused by Vanier and went on to state that Vanier had engaged in "manipulative sexual relationships" that had taken place under "coercive conditions" between 1970 and 2005. As Michael Coren pointed out in a column in the *Toronto Star* ("Vanier news is proof no one is above suspicion," *Toronto Star*, February 26, 2020, A13), "Here is a man who evinced love for the most profoundly challenged members of society, and then used and exploited needy women for his sexual gratification. It won't do to use platitudes here, and it's even worse to try to excuse the man because of the good that he did. That good, those works, created a reputation that then enabled him to hurt and to get away with it."

Vanier was, of course, not the only Canadian to be involved in attempts to help disabled people throughout the world. There are many others. Some of the most important of these contributions are documented in a book written by Henry Enns and Alfred H. Newfeldt entitled *In Pursuit of Equal Participation: Canada and Disability at Home and Abroad*. Canadians have been active in this field ever since the first National Conference on Rehabilitation of the Physically Disabled was held in Canada in 1951. Not only was this Conference responsible for bringing the Federal-Provincial Vocational Rehabilitation of Disabled

Persons Program into existence, and following this, the passage of the Vocational Rehabilitation of Disabled Persons Act of 1961, but also it provided a real stimulus to the development of many organizations for the disabled in Canada, including the Council of Canadians with Disabilities (CCD), the Canadian Association of Community Living (CACL), the Canadian Rehabilitation Council for the Disabled (CRCD), and others.

Through these and other organizations, as well as the concerted efforts of disabled people such as Gustave Gingras, Walter Dinsdale, Marie Barile, Allan Roeher, John Gibbons Counsell, Henry Enns, Jim Derkson, and many others, Canada and Canadians have been real pioneers in the development of many creative approaches to assisting people with developmental disabilities around the world, especially the community-based approach and cooperation between the public sector and the private sector in this rapidly expanding and extremely important field. The country also played a seminal role in the development of international organizations and activities in this area, and was the driving force behind the creation of Disabled People's International (DPI), the global non-governmental organization that has its origins and headquarters in Canada.

Much of the impetus and inspiration for this came from Henry Enns. He has been described as a pioneer, activist, and visionary for people with disabilities, and was a tireless leader at the local, regional, national, and international level throughout his life. He contracted rheumatoid arthritis when he was 15 years old, and by the time he was 19 he was in a wheelchair. Not only was Enns instrumental in creating the Canadian Centre on Disability Studies (CCDS), the establishment of many programs for the disabled in Manitoba and especially at its universities, and the creation of the Royal Bank Research Chair in Disability

Studies at CCDS, but also he was active in countless international organizations, programs, and projects, especially in Russia and the Ukraine. He also played an important role in the formation of Disabled Peoples' International (DPI) in 1980–81, and during the United Nations Year of the Disabled in 1981 assisted this organization in gaining a significant presence in the world. He travelled extensively in many different parts of the world despite his disabilities, visited more than 80 countries during the UN Decade of Disabled Persons between 1982 and 1992, and was President and Executive Director of DPI for several years. He was also active in creating the Disability Information Network, numerous on-line services, and many research facilities and resources for people with disabilities. He died in Sri Lanka while he was working on a Canadian Centre on Disability Studies project there for the Asian Development Bank.

Here is how Henry Enns and his contributions to the development of numerous programs and organizations for the disabled at the international level are described in a special tribute paid to him on the DPI website:

> In 1980, Henry saw the need for an international disabled rights organization that was run by disabled individuals for disabled individuals and was consequently elected Chair of the Steering Committee to start such an international organization at the Rehabilitation International Conference in Winnipeg. This led to the creation of Disabled Peoples' International (DPI) in Singapore in 1981. DPI was charged with the mandate of attaining and sustaining the rights of full participation and equality for disabled persons worldwide. Henry served as DPI's Deputy Chairperson, Chairperson and

Executive Director at different times throughout the 1980s and early 1990s. His vision for and dedication to DPI resulted in the far reaching organization that exists today.

Henry's strong vision of how things should be led to his significant contribution to the drafting of the UN World Program of Action concerning Disabled Persons. This contribution was later invaluable in the drafting of the United Nations Convention on the Rights of Persons with Disabilities (CRPD), which came into force in 2008 and has been signed by 146 nations. Henry and DPI were awarded the United Nations Testimonial Award in 1989 in recognition of this work.

Like Henry Enns, Terry Fox is another Canadian advocate and activist who had a remarkable impact on the world and the way disabled people are seen and treated throughout the world.

Terry was a superb athlete. He played soccer, rugby, and baseball when he was young, but his real passion was basketball. He went to Simon Fraser University at the urging of his mother, tried out for the junior basketball team, and earned a spot on the team ahead of many more talented players due to his fierce determination and perseverance. Unfortunately, he was involved in a severe car crash when he was young, an event that he believed contributed to the cancer that was discovered in one of his knees, although a number of doctors felt there was no connection between the two.

Fox was studying kinesiology at university in 1977 when it was ascertained that he had a rare form of bone cancer called osteogenic sarcoma, which led to the amputation of the better part of one of his legs. While he was recovering from this incredible ordeal, a creative idea hit him. Why not run a "Marathon of Hope" to raise money for cancer research? This would require running across Canada from coast to coast.

On October 15, 1979, Fox sent a letter to the Canadian Cancer Society announcing that he would like to undertake this run and appealing to the Society for financial assistance. He made no promise that his efforts would result in a cure for cancer, but he ended his letter with the following emotional appeal: "We need your help. The people in cancer clinics all over the world need people who believe in miracles. I am not a dreamer, and I am not saying that this will initiate any kind of definitive answer or cure to cancer. I believe in miracles. I need to."

The Canadian Cancer Society agreed to support Fox if he could find enough sponsors to support his run. He was successful in doing this. The Ford Motor Company donated a camper van; Imperial Oil donated the fuel; and Adidas supplied the running shoes, which was not an insignificant contribution in view of the number of running shoes that were required to undertake a run of this type and length. But the largest sponsor of all was a person who took up Fox's cause after the run began. It was Isadore Sharp, founder and CEO of Four Seasons Hotels and Resorts. He pledged to give two dollars a mile to the Marathon of Hope and persuaded close to 1,000 other corporations to do likewise after Terry Fox expressed his disappointment with the number of donations that came in during the early part of his run.

Fox commenced his Marathon of Hope in Saint John's, Newfoundland on April 12, 1980. With an unbelievable amount of courage and conviction, he ran through the Atlantic provinces, Quebec, and the larger part of Ontario before he was compelled to give up his run in Thunder Bay on September 1 when it was discovered that the cancer had returned and spread to his lungs. He had run 5,373 kilometres over 143 days and raised more than $1.7 million for cancer research when this happened. The pain must have been excruciating and unbearable. Canadians' awareness of Fox's efforts led to his being named Newsmaker of the Year in 1980 and again in 1981.

While the final tally of $1.7 million fell far short of Terry's ambitious goal—his objective was to raise one dollar for every Canadian or about $30 million in total at that time—this changed shortly after his run when the CTV Television Network organized a nationwide telethon in support of Fox and the Canadian Cancer Society. It raised $10.5 million. Donations continued to pour in during the winter, and by the following spring, a great deal more money had been raised. The Terry Fox Run became an annual event in Canada after that; 300,000 people took part in the first such run, raising $3.5 million.

Canadians have been so utterly taken with Terry's persistence, courage, and determination that they have simply never forgotten him. He epitomizes everything that is best in the Canadian character, and everything many Canadians aspire to and admire most. Douglas Coupland, who has done so much to keep Fox's memory and accomplishments alive through the various images he has created of Fox and the book he wrote about him, captured Terry's heroic spirit best in the summer of 2003 in an article for *Maclean's* magazine when he said: "There is not a soul in the land who could feel anything but pride towards the man's memory. How could they not? Terry loved his family and country, and he knew that as a people we have huge untapped reservoirs of kindness and strength."

While many Canadian disabled activists and advocates have made their presence felt in Canada and other parts of the world, none is more revered than Terry Fox. When Canadians were asked to pick the person they felt was the greatest Canadian of all time in a poll conducted by the CBC in 2004, Tommy Douglas, the founder of Medicare, placed first and Terry Fox second. Terry Fox has inspired millions of Canadians and people in other parts of the world with his vision, dedication, and commitment to a cause.

But the biggest tribute that could be paid to Terry Fox and his remarkable Marathon of Hope is what has happened in Canada and throughout the world since Terry passed away some three decades ago. The Terry Fox Run involves millions of participants in Canada and other parts of the world and has raised more than $750 million (Canadian) for cancer research as of January, 2018. In 2010, for instance, the 30th anniversary of the Marathon of Hope, the Run involved 1.9 million participants in 27 countries. Organized by Canadian embassies and consulates, Canadian Armed Forces bases, anti-cancer councils and societies, and international schools in many different parts of the world, the run is now one of the biggest one-day fundraisers for cancer research in the world. Virtually all the money raised outside Canada is retained by the participant countries and used to fund cancer research projects at recognized centres approved by the Terry Fox Foundation and the Terry Fox Research Institute. This is without doubt the finest honour that could be paid to Terry Fox by people and countries in all parts of the world today.

Like Terry Fox, Rick Hansen is a superb athlete and a person with disabilities. He was born in 1957 in Port Alberni, British Columbia. When he was young, he won all-star awards in five sports—an indication of the remarkable athletic prowess he has demonstrated throughout his life. Unfortunately, at age

15 Hansen was in a serious car accident that paralyzed him from the waist down and left him a paraplegic in a wheelchair. Despite this, he was the first student with a physical disability to graduate from physical education at the University of British Columbia. While there, he won national championships in wheelchair volleyball and basketball, and went on to become a world class champion marathoner and Paralympic athlete. He competed in wheelchair racing at both the 1980 and 1984 Summer Paralympics, winning a total of three gold, two silver, and one bronze medals, as well as 19 international wheelchair marathons and three world championships. He was and is an outstanding athlete who has won numerous national and international competitions and awards. He has also been a frequent commentator on radio and television with respect to the problems encountered by disabled people, most notably during the Olympic Games in Vancouver in 2010.

But this is not what Hansen is best known for in Canada and around the world. In 1985, Hansen decided to embark on a Man in Motion World Tour that would start in Vancouver to raise money for spinal cord research. The Tour attracted a great deal of international attention during his 26-month trek across 34 countries on four continents, logging more than 40,000 kilometres. The Tour raised more than $26 million for spinal cord research. Since that time, Hansen founded and acted as President and CEO of the Rick Hansen Foundation, which is devoted to making contributions to, and instituting programs for, spinal cord research. The Foundation has created and coordinates a major School Program and an Ambassador Program. It also provides an Accessible Cities Award that has been won by Winnipeg, Richmond, Edmonton, Mississauga, Regina, and Ajax, as well as established the Blusson Spiral Cord

Centre and the Rick Hansen Institute.

Married with four children, Hansen is author of several books, a member of Canada's Sports Hall of Fame and the Canada Walk of Fame, and has raised more than $200 million for spinal cord research and other socially worthwhile causes throughout Canada and the world. And this is not all. He is also an environmental activist, as well as the driving force behind the creation of an information network that is designed to track and record best practices in spinal cord treatment in Canada and other parts of the world. To this should be added the Rick Hansen Spinal Cord Injury Registry, which enables doctors and experts to share vital information on what works and what does not for people with specific types of spinal cord difficulties in all parts of the world.

As a result of Hansen's vision, dedication, and creative efforts, Canada is a world leader in spinal cord injury (SCI) research and innovation today. Starting as a modest initiative at the Vancouver General Hospital and the University of British Columbia, ICORD—the International Collaboration on Repair Discoveries—has become one of the largest interdisciplinary research programs in the world in this field, with many principal investigators and numerous trainees.

In addition, the Rick Hansen Foundation announced in 2019 that it was starting an exciting new project called Everyone Everywhere. As stated in its literature, "Everyone Everywhere means maximizing the potential of all though physical freedom. It means we're going to open the world and bring access to people in more places. It means we're going to rally an entire nation. And we won't stop until everyone can go everywhere" (#Everyone Everywhere).

The fact that Canada now has the Canadian Foundation for

Physically Disabled Persons and a Canadian Disability Hall of Fame is also due in no small part to the efforts and contributions of disabled athletes like Rick Hansen as well as Terry Fox and many others. Founded in 1987, this Hall of Fame inducts outstanding disabled athletes in sports such as alpine skiing, Nordic skiing, ice sledge hockey, wheelchair basketball and curling, as well as para snowboards. To this should be added the Winter Paralympic Games, where Canada and Canadians have played an important role. In 2010 in Vancouver, Canadian Paralympic athletes won 19 medals including 10 gold medals, and at the 2018 Games in Pyeonchang, South Korea, Canada's delegation of 55 Paralympic athletes won 28 medals, including eight gold, coming in second after the United States in the overall standings. It is achievements like this that account for the fact that Canada also played an important role in the development of the Invictus Games initiated by Prince Harry for wounded and disabled soldiers, which were held in Toronto from September 23 to 30, 2017 and involved 550 athletes from 17 countries competing in 12 events.

If Canadians have played an activist and highly creative role in improving the conditions and prospects for people with various types of disabilities throughout the world, they have also played a similar role in helping people suffering from many types of economic, social, educational, and medical difficulties and inequalities. One of the most important and successful developments in this respect in an historical sense was the Antigonish Movement. It started in Nova Scotia in general and at Saint Francis Xavier University in particular in the 1920s and '30s. It then spread to other provinces in Canada in the 1940s and blossomed internationally in the years to follow.

The key player in this development was Moses Coady, a

Catholic priest who became concerned about and committed to the welfare and well-being of the poor and less fortunate people of Nova Scotia. Relying heavily on inspiration provided by Father J. Tompkins and Hugh MacPherson, as well as the organizing abilities of his assistant, A. B. Macdonald, Coady was instrumental in creating the Extension Department at Saint Francis University in 1928 which established a series of economic and social programs that laid the foundation for the Antigonish Movement. This movement focused on adult education as a means of improving the economic and social circumstances of people by enabling trained organizers to go into communities in Nova Scotia to conduct a series of public meetings to assess the communities' strengths and weaknesses. This resulted in the creation of numerous study groups which often led to the formation of credit unions and cooperatives to sell fish, consumer goods, build homes, and market agricultural and industrial products.

These activities were based on a number of fundamental principles underlying the Movement, including the primacy of the individual, social reforms through education, focusing on economic advancements in communities first and foremost, group action rather than individual action, basic changes in economic and social methods, techniques, and organizations, and a full and abundant life for all people. The ultimate purpose of both the Extension Department and the Movement according to Coady was to create a more just and equitable world for all people in the present and the future. The title of Coady's book—*Masters of Their Own Destiny*—says a great deal about the central objective in Coady's career and in his life generally. It documents in detail the development of the Antigonish Movement and has been translated into numerous languages, thereby helping thousands of people throughout the world to improve their socio-economic status and situation.

During the 1940s, information about the Movement's successes spread rapidly throughout the world. By the 1950s, adult educators and social activists from Europe, Latin America, Africa, Asia, and elsewhere in the world were coming to Antigonish in large numbers to study at Saint Francis Xavier University, thereby leading to the establishment of the Coady International Institute at the university in 1959. Since that time, thousands of students have graduated from the university's highly respected leadership programs and returned to their own communities in more than 130 countries in the world to apply the methods and techniques learned at the Institute. This has resulted in the creation of myriad self-help groups of 10 to 20 members, especially in India and elsewhere in the world with an estimated 110 million members in total, as well as village savings and loan associations that have become very popular in Africa with over 10 million members. These groups and associations have become particularly helpful to women, largely because of the barriers confronting women in joining existing organizations as well as by increasing their collective bargaining power and producing numerous economies of scale and providing financial services where the objective is to achieve sustainable development at the community or local level.

Moses Coady was not the only Canadian activist to be concerned with the difficulties, inequalities, and problems confronting people. Stephen Lewis is another. He was born in Ottawa in 1937, son of David Lewis, former leader of the CCF (Co-operative Commonwealth Federation), the predecessor to the NDP (New Democratic Party). Stephen was schooled in politics from a very early age, and spent a year travelling in Africa in the 1960s, which he said transformed his life. He was the leader of the New Democratic Party in Ontario from 1970 to 1978, worked for several years as a labour mediator, columnist, and broadcaster in the early 1980s, and was Canada's ambassador to the United Nations from 1984 to 1988. It was during his first year in this position that Lewis worked closely with Brian Mulroney, Canada's prime minister at the time, as well as with the government of Canada in spearheading an international rescue effort during the brutal Ethiopian famine in 1984 that became "part of the greatest single humanitarian relief mission in history," in the words of Tony Burman, one of Canada's most respected international journalists.

From that point on, Lewis's life increasingly took on a more international character and orientation. From 2001 to 2006, he worked as the United Nations Special Envoy for HIV/AIDS in Africa. During this time, he was very outspoken about the lack of response from western governments and nations to the HIV/AIDS crisis, as well as the gap that existed between what they promised to give financially to the crisis and what they actually delivered.

Lewis's contributions in recent years have come primarily through the creation, administration, and chairing of the Stephen Lewis Foundation, which was established in the early years of the twenty-first century to "put money directly into the hands of community-based organizations working on the frontiers of the AIDS pandemic in Africa." From 2003 when the Foundation was created until June 2017, the Foundation distributed and committed over $106.2 million to programs and supported over 1,600 initiatives with 315 community-based organizations in 15 African countries. Its funding has been confined largely to the countries hardest hit by the pandemic, most notably Botswana,

Ethiopia, Lesotho, Malawi, Rwanda, Uganda, and Zambia. Its activities are focused on home-based care, grandmothers who are often required to look after AIDS victims and patients, orphan care, feeding programs, and art therapy. The Foundation is recognized for its innovative, sophisticated, and impactful approaches to programming in the field. In so doing, it epitomizes Canada's creative contributions to medical, health care, and healing problems as well as humanitarian causes and concerns through helping people in Africa and other parts of the world.

Stephen Lewis is not the only Canadian to have been an activist when it comes to addressing tragic situations in Africa. Roméo Dallaire is another.

Dallaire was born in Denelcamp in the Netherlands to a non-commissioned officer in the Canadian Army and a Dutch nurse. He spent a good part of his life in the Canadian Armed Forces and eventually served as Force Commander of the ill-fated peacekeeping mission of the United Nations in Rwanda—formally called the United Nations Assistance Mission for Rwanda or UNAMIR—between 1993 and 1994 that attempted to stop the genocide that was being waged by Hutu extremists. The genocide lasted a hundred days and led to the murder of some 800,000 Tutsies, Hutu moderates, and Twa, as well as the displacement of over two million people before this tragic and brutal affair was brought to an end. This experience profoundly affected Dallaire's life and directly and indirectly led to his involvement in a number of activist and advocacy activities and social causes and concerns in the years to follow, most notably dealing with genocide and its aftermath, experiences with post-traumatic stress disorder (PTSD), and attempts to end the involvement of children in wars, genocide, and other violent and vicious situations. Fortunately, Dallaire's experiences in these causes are documented in detail in the many books he wrote on these subjects in general and three books in particular.

In the first book, *Shake Hands with the Devil: The Failure of Humanity in Rwanda*, which was published in 2003, the genocide is described in great detail. The following year, Dallaire appeared before the International Criminal Tribunal for Rwanda to testify against Colonel Théoneste Bagosora. His appearance was of crucial importance in the overall outcome of the trial and having Bagosora convicted of genocide. Dallaire admitted that witnessing the genocide and playing a key role in trying to stop it has profoundly remained with him in later life, resulting in struggles with post-traumatic stress disorder (PTSD) and an attempt to commit suicide in 2000 when excessive drinking combined with taking an anti-depressant medication left him in a comatose state. This experience, and his attempt to act as an activist in bringing this terrible illness to the attention of armed forces personnel and the general public, resulted in a second book, *Waiting for First Light: My Ongoing Battle with PTSD*. During these difficulties, Dallaire was constantly haunted by the problem of compelling children to engage in genocide and participating in various battles and wars, so much so that he worked for a time as a special adviser to the Canadian Government on war-affected children and the prohibition of small arms distribution and child labour. In 2010, this experience led to the publication of a third book—*They Fight Like Soldiers, They Die Like Children: The Global Quest to Eradicate the Use of Child Soldiers*—the title of which speaks volumes about the frequency and tragedy of this situation.

These activist and advocacy activities also resulted in the making of a number of movies and television programs about

Dallaire and his experiences dealing with genocide, conflict resolution, humanitarian assistance, PTSD, and respect for human rights, especially as this affects soldiers, children, young people, and citizens in battle zones and war-infested areas and countries. They have also led to the conferring of numerous honorary degrees and the winning of many international awards for Dallaire's commitment and sacrifices in these areas and their inclusion in military developments and educational activities in different parts of the world.

This is not the only area where Canadians have manifested highly creative abilities of an activist and advocacy nature related to children and young people. Craig and Marc Kielburger have also achieved international reputations for their activism and advocacy in this area, largely as a result of their desire to improve the well-being of and opportunities for millions of children and young people around the world.

Craig Kielburger's life was transformed at a very early age when he read a newspaper article about a 12-year-old boy slave named Iqbal Masih in Islamabad, Pakistan who was murdered. Shortly after this, Craig and some of his high school friends created a non-profit organization called Free the Children. It has become one of the largest and most successful organizations for "children helping children" through education and social development in the world, having built myriad schools and school rooms and implemented hundreds of projects in Africa, Asia, Latin America, and the Caribbean since its founding in 1995. The organization began to receive international attention and recognition when it partnered with Oprah Winfrey's Angel Network and Craig Kielburger appeared on the *Oprah Winfrey Show* on several occasions.

Craig Kielburger is not only an accomplished social rights

activist and advocate. He is also an award-winning author and popular speaker who has been nominated for the Nobel Peace Prize on several occasions. Among his most recent books is *Me to We: Finding Meaning in a Material World,* which he co-authored with his brother Marc. It is a passionate plea for more kindness and understanding in the world. As the title suggests, the book is about making the shift from a world based on individualism, egoism, materialism, and concern for the self to a world based on community, caring, sharing, and concern for others. The book features articles by Oprah Winfrey, Richard Gere, Jane Goodall, Desmond Tutu, and others.

Me to We is not only a book. It also provided the impetus to bring about a major expansion in the operations and activities of the Kielburgers. In domestic terms, this resulted in the development of We Schools in Canada, the United States, and the United Kingdom which nurture compassion in students, empower them, and provide them with the tools they need to create transformative social change over the course of their academic studies. In 2016, 3.4 million students at 12,300 schools and groups across North America and in the UK were part of We Schools. In total, 6,940 organizations were supported, 2.2 million pounds of goods were collected, and $17.1 million was raised for these purposes. This resulted in the creation of We Villages that use a holistic sustainable development model to empower rural or marginalized villages in eight countries: Haiti, Nicaragua, Ecuador, Sierra Leone, Kenya, Tanzania, India, and China. The model is based on five pillars of impact: education, access to fresh water, health care, food sources, and economic and social opportunities.

And this is not all. Craig and Marc Kielburger have visited more than 50 countries in the world and been actively involved in the organization of many We Day Rallies, which have attracted thousands of students and young people to stadiums and arenas in different parts of the world. In 2017, for instance, the We Day Rally held at the Air Canada Centre in Toronto on September 28 featured motivational speeches by such well-known figures as Prince Harry, Ban Ki-Moon, Andre DeGrasse, Penny Oleksiak, Mia Farrow, and others. Tens of thousands of students attended the live event and it involved more than 1,200 schools in Toronto and vicinity, with millions more watching it on television.

In recent years, the multifarious activities of Free the Children, We Rallies, We Schools, and We Villages have been consolidated under a single organization called We Charity. In 2016, it received donations of $54.3 million and in 2017 completed building the We Learning Centre in Toronto that houses all of its operations and provides a more effective base for expanding its activities in Canada and throughout the world in the future. The organization's efforts are based on the conviction that when students and people act together they can earn a living, accomplish great things, find meaning in their lives, and change the world.[9]

Quite recently, the Kielburgers have embarked on another major undertaking to celebrate their twenty-fifth anniversary. It involves creating a 41,000-square-foot We Social Entrepreneurship Centre situated beside the learning centre in downtown Toronto. The purpose of the new centre is to provide opportunities to potential and promising social entrepreneurs to develop programs and projects that advance social causes and concerns across Canada and around the world, as well as train these people in the business, financial, and fund-raising

9. Craig Kielburger, Holly Branson, and Marc Kielburger, *WEconomy: You Can Find Meaning, Make a Living, and Change the World* (Toronto: Wiley, 2018).

practices they will require to develop and administer these programs and projects. In this case, the participants will be older—under 35—but still very committed to fulfilling the needs of young people in all parts of the world. It is a highly inventive initiative, but if it is as successful as the undertakings they have developed over their first 25 years, it promises to have an important impact on the world.

Activism and advocacy in environmental issues is another area where Canada and Canadians have made some highly creative and extremely important contributions to the world. Most of these contributions have come from people who have been very passionate about the natural environment and committed to finding viable solutions to environmental and ecological problems. In earlier times, there were the activist and advocacy activities of Jack Miner, Archibald Belaney, Ernest Thompson Seton, and Farley Mowat. As with Canada's earlier developments in popular music, the literary arts, and many other areas, these individuals created a strong historical base for the development of more recent environmental activists and advocates such as Maurice Strong, David Suzuki, Maude Barlow, the founders of Greenpeace International, and others.

What John Ruskin, William Morris, and Edward Carpenter did for environmental conservation in England and Henry David Thoreau, John Muir, Ansel Adams, and the Sierra Club did in the United States, Jack Miner, Archibald Belaney, and Ernest Thompson Seton achieved in Canada. Over time, the contributions of these environmental pioneers led to important developments in Canada and other parts of the world.

Jack Miner was born in Dover Center, Ohio in 1865 and came to Canada with his family in 1878. Like Belaney and Seton, Miner acquired a popular environmental nickname as a result of his commitment to nature and environmental preservation. He was called "Wild Goose Jack," and is deemed by some to be "the father of North American conservationism" for the highly creative role he played in the preservation of the natural environment in general and wildlife in particular.

It all began when Miner created a pond on his farm in 1904 and clipped seven tame Canada geese in the hope that they would attract wild geese. While it took time to realize this, wild geese and ducks eventually started to arrive at Miner's pond in large numbers in 1911. By 1913, his entire homestead had become a bird sanctuary, so much so that the Ontario government decided to help fund it. His practice of putting bands on geese and ducks began in 1909 when he banded what is believed to be the first duck in the world with an aluminum band. It was recovered five months later in South Carolina, thereby providing valuable information about the migration of ducks. Within the next few years, thousands of birds were banded in North America, thereby helping to create the United States Migratory Bird Treaty Act of 1918. The treaty made it illegal to capture, sell, or kill certain types of migratory birds.

In 1910, Miner embarked on a lifelong undertaking which involved speaking in public about wildlife conservation and the need to establish sanctuaries and wildlife refuges throughout the world. By promoting the formation of bird clubs, the building of bird houses, and commenting on the steady ecological deterioration of the Great Lakes, Miner became an extremely popular speaker on environmental matters as well as an activist for wildlife preservation. His world-famous bird sanctuary—the Jack Miner Migratory Bird Sanctuary—still exists in Kingsville, Ontario near Point Pelee National Park where he spent the bulk of his time. Presented with the Order of the British Empire

(OBE) by King George VI in 1943 for "the greatest achievement in conservation in the British Empire," Miner was rated by several American newspapers as one of North America's best known and most popular individuals, along with Henry Ford, Thomas Edison, Charles Lindbergh, and Eddie Rickenbacker. He was admired in Canada and throughout the world for the sanctuaries he created for geese, ducks, quail, pheasants, and other birds, as well as for banding over 50,000 ducks and 40,000 geese during his lifetime.

Despite Miner's remarkable contribution to environmental and wildlife conservation, Canada's best-known environmental activist and advocate in an historical sense is undoubtedly Archibald Belaney. Like Miner, Belaney was born in another part of the world, in this case Hastings, England in 1888. He came to Canada in his late teens and claimed to be of mixed Indigenous and European descent, as well as claiming that he was actually born in Mexico and not England. He was very enamoured of Canada's vast and beautiful natural environment, and eventually acquired the name "Grey Owl" after arriving in Canada. He married an Ojibwa woman named Angele Egwuna, assumed a First Nations identity and dress, and worked as a trapper, wilderness guide, and forest ranger in different parts of the country, such as Temagami in Northern Ontario.

Belaney separated from his first wife and eventually married a 19-year-old Mohawk Iroquois woman named Gertrude Bernard or Anahareo (Pony). She had a very strong influence on his life and encouraged him to stop trapping and start publishing his writings about the wilderness. Initially, this led to the publication of a number of articles in such periodicals as the famous English sporting and society magazine *Country Life*, as well as in the Canadian Forestry Association's publication

Forest and Outdoors. In addition to this, a film was made by Canada's National Parks Service called *Beaver People* that showed Belaney and his wife with two beavers they had taken in as infants and raised after their mother was killed. Following this, he was invited to join the Dominion Parks Service as a naturalist and moved with his wife to a cabin in Riding Mountain National Park for a time before going to Agawaan Lake in a home provided by the government at Prince Albert National Park, where Belaney was made honorary warden responsible for protecting beavers.

By the mid-1930s, Belaney was so popular as an environmentalist and speaker as well as one of Canada's and the world's first real environmental activists that he was invited to travel across Canada and go to England between 1935 and 1937 to promote his writings and speak about conservation and the natural environment. He was extremely popular in his Ojibwa dress and his lectures and first book—*Pilgrims in the Wild*—garnered huge audiences and sold more than 5,000 copies a month. In fact, he became so popular that he was invited to the British court where he made a presentation to King George V as well as Princesses Elizabeth and Margaret, the former of course now being the Queen. Two of the attendees at one of his lectures in England were David and Richard Attenborough. This eventually led to the film *Grey Owl* made by Richard (now Lord) Attenborough in 1999. It starred Pierce Brosnan, who also played James Bond in a number of films.

Belaney was haunted by problems throughout his life as a result of taking on a First Nations identity and appearance when he wasn't actually an Indigenous person, as well as by claiming that he was a descendant of an Apache tribe in the United States. However, he made numerous contributions to the environmental movement and conservation as an activist and advocate as well as the author of several books, including *The Men of the Last Frontier*, *Pilgrims of the Wild*, *The Adventures of Sajo and Her Beaver People*, and *Tales of an Empty Cabin*, in addition to inspiring a number of books about him including *The Tree* by his principal publisher, Lovat Dickson. Belaney was passionately concerned about nature and the need to protect it and the many animals that inhabited it throughout his life, as well as spreading the word about this to all parts of Canada, England, and eventually many other parts of the world.

Like Belaney, Ernest Thompson Seton was also born in another country, namely England in 1860. He was originally born Ernest Evan Thompson in Tyne, England to Scottish parents but due to a falling out he had with his abusive father he eventually changed his name to Ernest Thompson Seton, claiming that Seton was an important name in his paternal line and heritage. He was an outstanding wildlife illustrator and naturalist, often being called "Black Wolf" because of his fascination with wolves and especially wolves he acquired in Manitoba. He was appointed official naturalist to the Government of Manitoba in 1893, and, in 1907, undertook an epic 3,000-kilometre canoe trip across northern Canada.

Viewed by many as one of the principal founders of the Boy Scouts of America and a major influence on Lord Baden-Powell—founder of the Boy Scouts in England and generally throughout the world—Seton was a strong advocate for the Boy Scouts movement as well as Indigenous rights, despite the fact that he had a major dispute with some of the individuals involved in the scouting movement in the United States. He lived close to nature all his life, both in Canada and later in the United States. He became an American citizen in 1931 and settled in Santa Fe,

New Mexico where he was very active in the artistic and literary community in the 1930s and '40s.

More than anything else, Seton was a remarkable author, with more than 50 books to his credit dealing with such subjects as the natural environment, Indigenous peoples and their relationship with the land, various types of animals such as foxes, wolves, bears, and buffalo, the ways of life of wild animals, the woodcraft movement, and other matters. Among his most popular books were *Animal Heroes, Wild Animals I Have Known, The Biography of a Grizzly, The Gospel of the Redmen,* and *Woodcraft and Indian Lore: A Classic Guide from a Founding Father of the Boy Scouts of America.*

Much like Alexander Graham Bell, Seton is claimed by both Canadians and Americans as their own, as each man spent a great deal of time in both countries. Like Bell, Seton also emigrated to Canada when he was young and lived in Canada for many years before moving to the United States and eventually becoming an American citizen. It is not surprising, therefore, that both Canada and the United States have many parks and schools named after these two highly creative and very determined individuals, as well as numerous monuments that celebrate their creative achievements and preserve their memories.

The activist and advocacy activities of Miner, Belaney, and Seton, and especially the literary talents of Belaney and Seton, were carried on in a most convincing and compelling fashion by Farley Mowat. Born in 1921 in Belleville, Ontario, Farley Mowat's great-great uncle was Sir Oliver Mowat, a former premier of Ontario. His father was a librarian who fought at the famous Battle of Vimy Ridge during the First World War.

Farley became interested in animals, nature, and the natural environment when his family moved to Saskatoon when he was very young. He spent many long hours by himself exploring the surrounding countryside and studying the living patterns and characteristics of many different types of animals. These experiences played a crucial role in his life—as did the many trips he took and the lengthy times he spent in the far north living with Indigenous people—since they help to explain why he was such an outspoken and committed activist and agitator with respect to environmental issues and the need for conservation and change. They also help to explain why he wrote many popular books for people of all ages dealing with the circumstances, plight, and mistreatment of animals such as wolves, whales, seals, sea birds, and so forth. In total, he wrote more than 40 books that have sold more than 17 million copies worldwide and been translated into 52 languages. Particularly important are Mowat's books *Never Cry Wolf* (1963), which challenged and changed the negative stereotypes of wolves as vicious killers and predators; *People of the Deer* (1952), about the plight of Indigenous peoples in the Far North; *Rescue the Earth: Conversations* (1990), about his concern with the devastation of the natural environment and the need for environmental protection and preservation; and *Sea of Slaughter* (1984), about the impact water pollution, oil spills, and human activities generally have had on life in the Northwest Arctic over the last five hundred years. The title refers to the killing of countless whales, seals, fish, and species of sea birds. Mowat also wrote many internationally popular books for young people, including *The Dog Who Wouldn't Be* (1957), *Owls in the Family* (1961), and *Lost in the Barrens* (1956), which won the Governor-General's Award. Other works include *And No Birds Sang* (1979), about his experiences in the Second World War, and *A Whale for the Killing* (1972).

Despite the popularity of his work, Mowat was criticized by certain reviewers for fabricating some of the information contained in his books, particularly books having to do with the Indigenous peoples and tribes as well as his claims concerning animals and especially wolves. Mowat admitted freely and frequently that he occupied that grey zone between fact and fantasy, fiction and non-fiction, as have many other authors of books and makers of movies who have "stretched a point to make a point" in order to render their works more interesting to readers and viewers. However, what can be said in Mowat's defence is that he actually did spend a great deal of time in the Far North, in the wild, and on the seas collecting information and material for his varied and fascinating books about the interface between people, animals, and the natural world.

With historical precedents like these, it is no coincidence that Canada boasts several contemporary activists and advocates in the environmental field who have also had a remarkable impact. One of the most notable of these is Maurice Strong, who made many outstanding contributions to the modern environmental movement throughout the world during the course of his life. Strong worked tirelessly for more than 40 years to promote environmental causes, conservation, and consciousness. He was a forerunner in the development of the modern environmental movement in much the same way that William Osler was a forerunner in the development of modern medicine, largely because of the seminal role he played in the establishment of numerous environmental organizations, programs, initiatives, events, and conferences. It is for this reason that some authorities in the environmental field consider Strong to be "the founder of the contemporary environmental movement."

Strong was born during the Great Depression in Oak Lake, Manitoba. His remarkable work on environmental issues dates largely from the time he spent at the United Nations. His involvement in this organization got off to a rather inauspicious beginning when he was appointed to a low-level position as a junior security officer at the UN in Lake Success, New York in 1947. He left the organization shortly after this, but returned to the UN in 1971 and had the foresight to commission a report on the state of the planet at that time—*Only One Earth: The Care and Maintenance of a Small Planet*—which was co-authored by Barbara Ward and René Dubos, two of the world's most outstanding environmentalists and scholars of that era. The report summarized the findings of over a hundred and fifty leading experts on the environment from 58 different countries in preparation for the first-ever UN conference on the environment that took place in Stockholm in 1972. It was the first time a "state of the environment" assessment had been made. In conjunction with other developments, this led to the establishment of the United Nations Environmental Program (UNEP) in Nairobi, Kenya in 1972 by the General Assembly of the United Nations, and the election of Strong as Director of the aforementioned program. As head of the UNEP, Strong was responsible for convening the first-ever meeting of international experts on climate change and global warming.

In 1983, Strong was appointed to the World Commission on Environment and Development. Usually referred to as the Bruntland Commission after its chair Gro Bruntland, former president of Norway, the commission was an independent body established by the United Nations. Its executive director, James McNeil, who was also a Canadian, was responsible with Strong and the other commissioners for creating and promoting the idea of "sustainable development," which has had a powerful impact on the world ever since. It quickly became a term that was and still is used extensively throughout the world to connote development that takes the natural environment and future generations—and not just the present generation—into account in all developmental planning, policy, and decision-making. In many ways, this idea in general, and the Bruntland report in particular, are responsible for the fact that environmental assessments are now usually compulsory in all parts of the world when major developmental activities or construction projects are being considered, planned, and organized.

Strong's reputation at the United Nations and in the international environmental community led to his appointment as Secretary-General of the United Nations Conference on Environment and Development—the so-called Earth Summit—convened in Rio de Janeiro in 1992. Strong complained bitterly and publicly at this summit that the participants adopted many sound environmental principles and management practices but failed to commit the necessary financial and human resources to translate these principles and practices into reality and prevent a global environmental catastrophe. He was chastised by many governments and government officials for his comments on this matter, but stood his ground and was eventually supported by the Secretary General of the United Nations at that time, Boutros Boutros-Ghali.

After the summit, Strong continued to play a leadership role and act as an activist and advocate for environmental change and progress throughout the world, largely by implementing several of the key agreements concluded at the Earth Summit through the creation of the Earth Council, the Earth Charter movement, chairmanship of the World Resources Institute, and his membership on the Board of the International Institute for Sustainable Development, the Stockholm Environment Institute, the Institute of Ecology in Indonesia, the Beijer Institute of the Royal Swedish Academy of Sciences, and others. This was complemented by his involvement in, and founding and/or administration of, many other international political, educational, environmental, and other organizations, such as the World Economic Forum, the World Bank, the Center for International Development at Harvard University, the World Business Council for Sustainable Development, the World Wildlife Fund, Resources of the Future, and others. Small wonder Strong was accorded more than 50 honorary doctorates from universities all over the world during his lifetime, as well as numerous international awards for his work in the international development field in general and the environmental field in particular.

While Strong was a key player on the world stage for more than 40 years, he was a source of controversy and confusion for many people in the international community because they were never quite sure where he stood on many important global issues and problems. This is because he was involved in so many business, corporate, and private sector activities in addition to his public sector activities and responsibilities that it was difficult to determine "where he was coming from," as they say. Was he acting as an environmentalist, humanist, activist, diplomat, businessman, corporate executive, or entrepreneur? Since it was hard to determine this, controversy haunted Strong throughout his life. Nevertheless, this should not be allowed to obscure the fact that he made many crucial contributions to the development of the modern environmental movement and global development. The world owes Strong a great debt of gratitude for the courage, determination, vision, and creativity he demonstrated in bringing many difficult environmental and developmental issues to a head and the many programs, institutions, and projects he created or was involved in creating that produced beneficial effects in all parts of the world.

Another Canadian environmental activist and advocate who has had a powerful impact on the world is David Suzuki. He was born in Vancouver in 1936 following the migration of his maternal and paternal grandparents to Canada from Japan early in the twentieth century. Although he is a third-generation Japanese-Canadian, he was held with his family in an internment camp in British Columbia during the Second World War. After the war, the Suzuki family was compelled to move around Canada a great deal in search of work because the government confiscated and sold their dry-cleaning business during the war.

After specializing in zoology and genetics at university, David Suzuki got a job teaching in the genetics department of the University of British Columbia where he remained for nearly 40 years. While his work on a variety of genetic issues and environmental problems is well-known in international academic circles, he is much better known throughout the world for his work in broadcasting in general and programs like *Suzuki on Science, Quirks and Quarks*, and *Science Magazine* in particular. His crowning achievement in broadcasting, however, is undoubtedly *The Nature of Things*. This well-known and highly acclaimed Canadian Broadcasting Corporation (CBC) television series premiered in 1960 and Suzuki became its host in 1979. The program has aired in more than 50 countries around the world. Through this program, Suzuki has made people more aware of the natural world and all things in it, as well as threats to the environment and the need for more sustainable development in the future.

A highly creative and well-respected environmental activist, advocate, and scholar, Suzuki has been outspoken on many issues related to climate change and the damage done to the natural environment and the earth as a result of humanity's obsession with materialism and economic growth. He has commented at length on blatant attempts to discredit climate change documentation, research, and researchers by political, governmental, corporate, and climate-change deniers. In his view, evidence on climate change from the Intergovernmental Panel on Climate Change and many distinguished scholars and experts in this field is undeniable, irrefutable, and beyond reproach, especially with regard to major fluctuations in weather patterns and climatic conditions, record temperatures, heat waves, forest fires, shrinking glaciers, rising sea levels, and other pertinent indicators and indications. Without major changes in the ways of life of people and their lifestyles—ways of life and lifestyles that are intimately connected to consumption, consumerism, the economy, and economic growth—the world will not be able to cope with all the environmental problems and ecological disasters that are already occurring in the world or looming up on the horizon. Suzuki has long argued that we are rapidly running out of time to address these problems and disasters.

Despite his busy schedule, Suzuki has found time to write more than 40 books on the environment, climate change, his Japanese heritage, and related matters, including *It's a Matter of Survival*, *A Time to Change*, *The Sacred Balance*, and *The Big Picture: Reflections on Science, Humanity, and a Quickly Changing Planet*, which he co-authored with David Taylor in

2009. In addition to this, he was instrumental along with his wife Dr. Tara Cullis in founding the David Suzuki Foundation in 1990. This Foundation is committed to "finding ways for society to live in harmony with the natural world that sustains us" through clean energy, measures to reduce climate change, and initiatives to ensure sustainable fishing, farming, forestry, and especially industry.

The Foundation's stated mission is "to protect the diversity of nature and our quality of life, now and for the future." It includes among its basic goals and objectives protecting our climate; transforming the economy; protecting nature; reconnecting with nature; and building community. Interestingly, he and his wife's daughter, Severn Cullis-Suzuki—who is also extremely active in the environmental movement—closed a plenary session at the 1992 Rio Earth Summit referred to earlier with a powerful and persuasive speech about the importance of the environment and the need to preserve and protect it to the battery of politicians who were in attendance at the Summit.

Yet another major Canadian forerunner and frontrunner in the international environmental movement is Maude Barlow. She was born in Canada in 1947 and learned a great deal about social justice and domestic and international activism at a very early age from her father, Bill McGrath. He was well-known in Ottawa for his involvement in fighting wartime atrocities and promoting reform of Canada's prison system when he returned home to Canada after the Second World War.

Early in her career, Barlow was very active in the feminist movement in Canada, largely as director of the Office of Equal Opportunity for the City of Ottawa and advisor to Pierre Elliott Trudeau's Liberal government on equality for women. Following this, she fought Brian Mulroney's Conservative government on the North American Free Trade Agreement (NAFTA) and then became an active participant in the quest to create a strong, independent, and sovereign Canada through her involvement as national chairperson of the Council of Canadians, a position she still holds in an honorary capacity today.

In recent years, Barlow's work has become much more international in nature. She has become extremely active in the grassroots movement to ensure that all citizens have sufficient access to water, as well as to prevent the depletion, pollution, diversion, and commodification of water by major corporations in Canada and other parts of the world. This has caused her to travel extensively throughout the world speaking about the present water crisis in the world, as well as to become actively engaged in the Blue Planet Project, an international civil society project that she and the Council of Canadians created to protect the world's fresh water from the escalating threat of international trade, privatization, exploitation, and exportation.

The Project works with organizations and activists in the South and North, and is affiliated with international networks such as Friends of the Earth International, Red Vida (the American Network on the Right to Water), the People's Health Movement, and others. The Project is also working with worldwide partners on evolving a human rights framework to protect water for people, nature, and future generations which is local, regional, and national as well as international in character. Growing out of concern over the fact that the world is running out of fresh, clean, and accessible water—not just in Africa, Asia, and Latin America but also in Europe and North America—the Project is involved in creating a movement to secure an international treaty on the Right to Water, as well as a commitment from the United Nations to fulfill this objective as one of the most important human rights of all, especially as this right was overlooked when the Universal Declaration of Human Rights was signed in 1948.

Fortunately, this objective was realized in 2010 when the United Nations General Assembly adopted an historic resolution that recognized the human right to safe and clean drinking water and basic sanitation. A second resolution was adopted by the Human Rights Council affirming that drinking water and sanitation are human rights. This established the responsibilities that all governments are expected to fulfill concerning these rights. It also clarifies the fact that the resolution passed by the General Assembly is now legally binding on countries under international law.

In recognition of these efforts, and others, Barlow was the recipient of the Right Livelihood Award in 2005—an award that is commonly regarded as "the alternative Nobel Prize" because it honours people "offering practical and exemplary answers to the most urgent challenges facing us today." She was also recognized with the Canadian Environment Award in 2008, the Planet in Focus Eco Hero Award in 2009, and the Earth Care Award in 2011—the highest international award given by the Sierra Club in the United States. She is also the author of many books on these matters, including *Take Back the Nation, Global Showdown: How the Activists are Fighting Global Corporate Rule, Too Close for Comfort: Canada's Future Within Fortress North America*, and especially *Blue Covenant: The Global Water Crisis and the Battle for the Right to Water* and, more recently, *Blue Future: Protecting Water for People and the Planet.* The titles of these books say a great deal about the type of creative activity and environmental activism Barlow has been involved in over many years. They also provide confirmation of the vital role she has played—and is playing—everywhere in the world through her involvement in organizations such as the Washington-based Food & Water Watch, the Hamburg-based World Future Council, the San Francisco-based International Forum on Globalization, and the United Nations where she has served as senior adviser on water issues.

There is another environmental activist—or perhaps I should say group of environmental activists because it involves several Canadians—that has had a major impact on the world in environmental terms. It is the Canadians who were involved in the founding and early development of Greenpeace and, in certain cases, Greenpeace International at a slightly later date. The group included Bob Hunter, David Fraser McTaggart, Bill Darnell, Paul Watson, and others. When their respective contributions are added up, they make a very convincing case about the creativity that was forthcoming from Canada and Canadians in bringing this unique organization into existence and eventually expanding it substantially throughout the world.

Bob Hunter was a journalist whom Greenpeace recognizes on its website as "a relentless visionary and a mystic storyteller ... who infused the young Greenpeace with a magic that lasts to the present day." It is largely as a result of Hunter's media experience and expertise that Greenpeace became a household name around the world as a result of projecting consciousness-changing sound bits, bites, and images—what Hunter called "media mind bombs"—throughout the world in the guise of news. Here is what Greenpeace has to say about Hunter and the role he played in the founding and early development of the organization and its well-known approach to environmental issues:

Combining creativity with strategic thinking and a hard-nosed journalistic sense for a good story, he helped to shape—perhaps like no other founding member—what would come to be known around the world, as a "*Greenpeace action.*" Hunter's sprit of courage, defiance and media-saviness continues to define Greenpeace up to the present day. The organisation he co-founded and shaped in a way few others have, will always be blessed with his spirit.

Like Hunter, David McTaggart also played a crucial role in this process but in a very different way. He took Greenpeace's "free-spirited founding ethos and translated it into an international organisation," according to authorities on this respected institution.

This commenced in 1972 when McTaggart took his personal boat, renamed it *Greenpeace III*, and sailed to the Pacific Ocean to protest the testing of nuclear weapons by the French Government. This became known as "classic Greenpeace"—a tiny boat challenging one of the mightiest military forces in the world, much like the solitary citizen who challenged a huge tank in Tiananmen Square in Beijing, China. McTaggart's protest proved very successful. In 1974, the French government announced that it was terminating its atmospheric nuclear testing program.

McTaggart's pragmatism and highly entrepreneurial approach in the minds of Greenpeace members was also responsible for building Greenpeace into an effective international organization, Greenpeace International, largely by melding the separate strands of the organization together to form a single, coherent entity. By 1985, the organization that started on a small fishing boat had three ships and 50 campaigns around the world under its belt. According to the organization, "The foundations for this were laid by the man who sailed off on a whim only to end up dedicating his entire life to environmental issues." McTaggart became chairman and chief spokesman of Greenpeace in 1979, and continued to remain active in the organization after his retirement until his untimely death in Italy in 2001 as a result of a car accident.

The third member of the group was Bill Darnell. While his role in Greenpeace was less significant than Hunter's or McTaggart's, he is given credit by a number of the founding members with naming the organization. This occurred when members of the organizing committee were leaving a church basement one day after a meeting and someone said the word "peace" as a farewell greeting. Bill Darnell immediately replied, "Let's make it a green peace." From there, the two words were linked together by the son of Jim Bohlen—another member of the organizing committee—to form the word Greenpeace for use on pins, badges, letterhead, and so forth.

The final member of the group is Paul Watson. He was a member of Greenpeace at the very beginning, but for a very short period of time when Greenpeace was just getting started and was involved in the protests against the American atomic tests in Alaska. However, Watson believed that Greenpeace's programs and protests were too mild and not sufficiently aggressive to prevent damage to the natural environment. Consequently, he left Greenpeace and formed his own organization called Sea Shepherd Conservation Society. It is based on the conviction that militant measures and "direct action" are imperative to prevent further destruction of the natural environment and damage to the world's ecosystems. Calling himself an "ecowarrior," Watson has gone to jail for confronting environmental polluters and practising his strong beliefs and commitment to environmental preservation and conservation, especially with respect to marine life.

Greenpeace has come a long way since 1971, when a handful of concerned citizens set sail from Vancouver in a hired fishing boat to protest the testing of nuclear weapons in the U.S. test zone of Amchitka. Greenpeace was successful in blocking this test and several others in the years to follow. It now operates in more than 40 countries, has more than 3 million members, and is committed to protecting biodiversity and preventing pollution everywhere in the world.

With creative Canadian environmental activists and advocates like these, and others that could be mentioned, it is easy to understand why people, institutions, and countries in other parts of the world have been disappointed with Canada's performance on environmental issues and the environmental front generally in recent years. Whereas they expected strong leadership and forceful action on the part of Canada, Canadians, and Canadian governments on these matters and problems, it is disconcerting to realize that Canada is still one of the largest polluters in the world, Canadians are producing one of the largest ecological footprints on the world, and the government of Canada dragged its feet on environmental actions and climate change for many years, especially with respect to such initiatives as the Kyoto Accord and the agreement hammered out in Copenhagen, Denmark in 2009. It was with some relief, therefore, that Canadians saw Canada play a more active and engaged role in the signing of the Paris Accord on climate change in 2016.

Given the present situation, however, a great deal of creativity and funding will obviously have to go into ensuring that Canada and Canadian citizens, corporations, and governments develop the plans, programs, policies, and legislation that are required to deal effectively with the country's and the world's environmental needs, problems and possibilities in the future. Anything less than this will not be equal to the challenge, which is growing rapidly in all parts of the world and threatening to escalate out of control.

There is one final Canadian activist who deserves a prominent place in this chapter because she has had a powerful effect on the world through her thoughts, beliefs, research, writing, and popular books. It is Naomi Klein, one of Canada's and the world's best-known activists.

Klein was born in Montreal in 1970 to a long line of social and political activists, especially her mother Bonnie Sherr Klein, who was a well-known documentary filmmaker, and her father, who was a physician and member of Physicians for Social Responsibility. Moreover, her husband, Avi Lewis, son of Stephen Lewis, is also a well-known activist, TV journalist, and filmmaker. Together, they form a dynamic couple, and are very active in the New Democratic Party in Canada.

What is most remarkable about Naomi Klein is the courage she has demonstrated in tackling some extremely contentious, controversial, and fundamental international issues at a very early age, especially the dominant role played by capitalism, capitalists and multinational corporations in global development and human affairs over the last century as well as at present. Klein made her position on this matter well known in her first major book—*No Logo: Taking Aim at the Brand Bullies*—which was published in 2000. It became very popular soon after it was published and acted as a lightning rod and manifesto for the anti-corporate globalization movement throughout the world. It was based on the belief that brand-based consumer culture was running rampant in the world and large multinational corporations were responsible for exploiting workers in the world's poorest countries in pursuit of ever greater profits. It quickly became an international bestseller with sales of more than a million copies in 28 different languages.

This belief was taken up in earnest and magnified considerably in another book—*The Shock Doctrine: The Rise of Disaster Capitalism*—which was published in 2007. By this time, Klein was ranked eleventh in a list of the world's top 100 public intellectuals compiled by *Prospect* magazine in conjunction with *Foreign Policy* magazine. In this book, Klein contended that free-market neoliberal policies—especially those espoused by Milton Friedman, the well-known economist and monetary theorist from the University of Chicago, and his followers—were embraced by some Western countries and predicated on a deliberate strategy of exploiting national crises in order to push through controversial plans, policies, and agendas when citizens were either too distraught or physically distracted by disasters or upheavals to mount effective resistances to them. She cited many examples of this, such as the terrorist attacks of 9/11 on the United States and the War on Iraq, the transformation of South American economies in the 1970s, and others. While this book generated a great deal of controversy among reviewers and critics who provided differing views and opinions about the central argument, it was another international best-seller and resulted in the making of a film documentary on this book.

Yet another book by Klein—*This Changes Everything: Capitalism vs. The Climate* published in 2014—extended many of the arguments in her first two books about the policies and practices of capitalism and multinational corporations into the environmental realm. Here, Klein contended that the status quo in the world is no longer a viable option due to climate change, global warming, and the environmental crisis, which are becoming antithetical to the realization of a better and more sustainable world. In Klein's view, it is necessary to redesign the global economic system, transform the world economy, reformulate current political policies, practices, and systems, and abandon the "free market" ideology of contemporary times since this is having a devastating effect on the planet and creating incredible injustices and inequalities in income and wealth throughout the world. Many of these problems, and the problems raised in her earlier books, were taken up in Klein's more recent book *No Is Not Enough: Resisting Trump's Shock Politics and Winning the World We Need*, which was published in 2017.

What is most exemplary about Klein's activism and advocacy—regardless of which side one comes down on with respect to many of her most pronounced arguments, ideas, and beliefs—is how meticulous she has been and is in researching the subjects she writes about as well as her ability to provide concrete examples of the various cases she is making. In these and other books as well as in many articles she has written, it is clear that she has an in-depth knowledge, understanding, and awareness of what is really going on in the world today, what the key issues and problems are, and how these issues and problems can be dealt with effectively in the future in Canada and other parts of the world. As such, she epitomizes the importance and impact that Canadian activists and advocates have had on the world over the last few centuries in many ways.

Chapter Seven
Sports and Recreation

If Canada has made many creative contributions to activism and advocacy that have had a powerful effect on the world, the same can be said for sports and recreation. Millions of people throughout the world are involved in or enjoy a variety of sport and recreational activities created or enhanced by Canadians.

As Susan Hughes points out in her book *Canada Invents*, "Canada is a country full of people who like to have fun. Canadians have adapted or changed games from other countries to suit their own special needs, and they have come up with their own completely original ideas about how to enjoy themselves."[10] In the process of doing this, they have transformed some sports and recreational activities that originated in other parts of the world to suit their own specific needs and circumstances, and have created many other activities in this area that are appreciated throughout the world.

Generally speaking, sports and recreation can be divided into two types of activities. On the one hand, there are activities that require some form of competition. This competition may be individual or collective, amateur or professional, mental or physical, but it involves contesting certain skills and abilities and awarding cups, trophies, medals, and championships to the winners. A great deal of creativity is manifested in sports like this, not only in terms of how athletes devise their training schedules and develop their strategies and tactics for competing and winning, but also in terms of how managers and coaches

10. Susan Hughes. *Canada Invents* (Toronto: Owl Books, Maple Tree Press, Inc., 2002), p. 31.

assess the talents and abilities of athletes and match them up to produce the best possible results. Most sports are designed, organized, and played in this way, such as hockey, basketball, baseball, football, lacrosse, curling, rowing, figure skating, skiing, and so forth. On the other hand, there are activities that do not require some form of competition or winning, such as gardening, camping, reading, hiking, watching television, travelling at home or abroad, visiting parks, museums, art galleries, and conservation areas, and other forms of recreational and leisure time activity.

While Canadians are involved in both types of activities, many enjoy activities that involve some form of competition. This is especially true for sports that are closely related to the country's climate, geography, and natural environment, such as hockey, curling, skiing, figure skating, speed skating, rowing, and so forth. This makes "place" an important factor in determining the specific character of Canadian sports and recreation. As S. F. Wise pointed out in talking about the innovative contributions Canadians have made to the creation and development of many sports over the centuries:

> Canada's sharply defined seasons, its bountiful water resources, both salt and fresh, and the demands its environment placed on such pioneering and survival virtues as strength, endurance and mental and physical toughness, all influenced the manner in which our sports developed. Moreover, organized sport was enormously important to Canadians for the relief it provided from the task of earning a living in a hard country. *To an extraordinary degree, therefore, Canadians have been among the world's most creative innovators in sports.*

Both basketball and modern ice hockey are Canadian in origin; both species of North American football owe a good deal to Canadian creativity. Lacrosse is a Canadian adaptation of an ... [Indigenous] game. Modern curling was revolutionized by Canadian-developed techniques; even such activities as ornamental swimming and competitive water-skiing are largely Canadian in origin.[11]

One cannot travel very far down the road to sports and recreation in Canada without encountering the incredible importance of hockey. It grips the Canadian imagination like no other sport or recreational activity. Canadians are obsessed with it, and watch it and play it all the time in numbers well into the millions. It is often said that Canada is a "hockey nation" first and foremost and that hockey is "Canada's game." This assertion is generally recognized throughout the world and attests to the remarkable impact hockey has had and continues to have on Canada, Canadians, and the world at large.

Since there are many different types of hockey—ice hockey, field hockey, roller hockey, ball hockey, and so forth—it is important to emphasize that it is hockey on ice we are discussing here. While the origins of this sport can be traced back to classical times, there is still an active debate over whether ice hockey was first played in Canada or England, since certain types of stick and ball games were played on ice in northern England in the early part of the nineteenth century. However, there is a great deal of evidence to support the contention that Canadians created ice hockey as a sport due to the seminal role they played in its development at the outset. Not only did they play a pioneering role in the creation of most of the equipment

11. S. F. Wise, "Canada's Sporting Tradition," in Charles J. Humber, Editor-in-Chief, *Canada: From Sea unto Sea*, p. 359 (italics mine).

that is used in the game today, such as various types of skates, pads, sticks, and goalie masks, but also they played a crucial role in setting out the rules and procedures that govern the game, from the number of players on each side and regulations with respect to passing, offsides, onsides, and face-offs to the number of periods, the penalty shot, overtime periods, the dreaded "shoot out," and a great deal more.

As far back as the 1850s, elementary versions of ice hockey were played in Halifax and Windsor in Nova Scotia and Kingston in Ontario, thereby making it difficult to ascertain precisely where this sport originated in Canada. By 1875, however, the game was played in many parts of the country, especially in Montreal where the first rules of the game were set out by W. F. Robertson, Dick Smith, Chick Murray, and especially J. G. A. Creighton, a student at McGill University who wrote the first rule book on hockey. The first organized hockey team in Canada was the McGill University Hockey Club. It was created in 1879 and won the first "world championship" in hockey played at the Montreal Ice Carnival in 1883. This is when the first national hockey association was created, the Amateur Hockey Association of Canada, which was followed by the creation of the Ontario Hockey Association in 1890.

Lord Stanley, Governor-General of Canada from 1888 to 1893, was a great fan of this game and donated the trophy that was awarded to the national champion and is still awarded to the National Hockey League (NHL) champion today. It was initially called the Dominion Hockey Challenge Cup but later the Stanley Cup in honour of its founder and donor. It was won for the first time by the Montreal Amateur Athletic Association in 1893, which won it again in 1894. This was fitting in view of the phenomenal role Montreal played in the development of hockey over the next 120 years, from the four Stanley Cups won by the Montreal Victorias from 1895 to 1898 to the numerous Stanley Cups won by the Montreal Canadiens since that time.

While hockey remained an amateur sport for many years, pressure built rapidly to professionalize it and expand it to other parts of North America. This began in 1908, when the Ontario Professional League was created. A year later, the National Hockey Association was formed, becoming the NHL, as it is still known today, in 1917. Interestingly, the International Ice Hockey Federation (IIHF) was also created about this time. Founded in Paris in 1908, it now has 74 members. This fact helps to explain why men's hockey has been played at the Winter Olympics since 1920, although women's hockey was not played at the Olympics until 1998. In both cases, however, Canada dominated the sport when it was played at the Olympics for the first few times, although competition is much tougher and keener today than it was originally.

As hockey became more popular and professional, it expanded rapidly in North America. Leagues were formed in both the east and the west of Canada, followed by the "original six" NHL teams in Montreal, Toronto, Detroit, Chicago, Boston, and New York. They competed for the Stanley Cup between 1942 and 1967, during which time the Montreal Canadiens and Toronto Maple Leafs dominated and won many Stanley cups.

In 1972, the World Hockey Association (WHA) was created to capitalize on the growing popularity of hockey in North America, as well as to challenge the supremacy of the NHL. Canada was well represented in the association with teams in Ottawa, Quebec City, Edmonton, and Winnipeg. However, the WHA began to experience financial difficulties soon after it was founded and had to be disbanded in 1979. This set the stage for a dramatic expansion of the NHL, giving rise to the 31 teams that compete for the Stanley Cup today, most of which are now located in the United States, with another team from that country to be added in Seattle in 2020 or 2021.

While growth in hockey was most evident in North America during the earlier part of this era, a number of developments were beginning to take place in Europe that were stimulating interest in hockey in that part of the world as well. Spurred on by the fact that the IIHF had its headquarters in Paris, hockey was played at the Winter Olympics in 1920 and every four years thereafter. When what is generally regarded as the oldest hockey tournament in the world was commenced in Davos, Switzerland in 1923 with the winner being awarded the Spengler Cup, hockey really rose in prominence and popularity in many European countries, especially Russia, Sweden, Finland, and Czechoslovakia in the beginning and France, Switzerland, Germany, and Latvia more recently.

It wasn't long before demands were made for competition between the best teams from North America and Europe through the creation of "world hockey championships." In the early years, and especially during the period from 1920 to 1961, amateur teams from Canada dominated international competitions like this. However, following this, the Soviet Union dominated international hockey for many years. Its teams won nine out of ten world amateur hockey championships at one time, although many claimed that its players were professionals and not amateurs because they spent the bulk of their time playing for the Red Army team when they were in the Soviet armed forces.

This set the stage for one of the most dramatic and exciting showdowns in the history of sports. It was a well organized and highly hyped "winner-take-all" series between Canada and the Soviet Union to determine which country had the best hockey team and players in the world. It took place in 1972 and was an eight-game series involving Team Canada and Team Soviet Union, with the first four games being played in Canada and the final four games played in the Soviet Union. Assuming that Team Canada would win the series easily, Canadians were stunned when the USSR won two of the first four games in Canada and tied another. When the series moved to the Soviet Union and the Soviet team won game five, Team Canada had its back to the wall. It had to win all the remaining games to win the series. And this is exactly what it did. After winning games six and seven, Foster Hewitt announced at 19:26 of the third period in the eighth and final game, "Here's another shot, right in front.... *They score! Henderson has scored for Canada.*"

The whole country rejoiced when this occurred. It was one of those rare moments in Canadian sports—indeed in the history of Canada and the history of sports throughout the world—that will simply never be forgotten. While a number of other feats have been achieved by Canada in international hockey since that time—including a game-winning goal by Mario Lemieux in a later Canada-Russia series and what has been described by many as "the golden goal" by Sidney Crosby at the Vancouver Winter Olympics in 2010—what made Henderson's goal in the final game of the 1972 Canada-Soviet series so special was the fact that Canada had regained its reputation in world hockey. It is a reputation that has been largely sustained until quite recently despite a few ups and downs, with Team Canada winning both the World Cup as well as the gold medal at the Olympic Games in Salt Lake City in 2006, Team Canada winning the gold medal at the Olympic Games in 2010 and again in 2014 in Sochi, Russia.

Canada's reputation in international hockey has been enhanced considerably by the remarkable achievements of Canadian women and the creation of Canada's women's hockey team as well as other women's hockey teams in the world. Here also, Canada and Canadians set the stage for many developments that happened later throughout the world.

Canadian women began their ascent to their long period of dominance in international hockey in the 1990s. While women have been playing hockey in Canada for more than a century—and have achieved numerous "firsts" and "successes" along the way—it is the ascent of the women's national hockey team to the apex of international hockey that was the most spectacular of all. The national team was clearly in a class by itself for a long time. Not only did it win world championships in women's hockey in 1990, 1992, 1994, 1997, 2000, 2001, 2003, and 2006, but also it won gold medals at the Olympic Games in 2002, 2006, and 2010, as well as a silver medal at the Olympic Games in Nagano, Japan in 1998 when women's hockey was played for the first time. This made stars such as Hayley Wickenheiser, Cassie Campbell, Danielle Goyette, Angela James, Vicky Sunohara, Jayna Hefford, Jennifer Botterill, Nancy Drolet, Kim St-Pierre, Gillian Apps, Cherie Piper, Shannon Szabados, Marie-Philip Poulin, and others household names in Canada. It also led to another cup named after a Canadian Governor-General—the Clarkson Cup named after Governor-General Adrienne Clarkson—being awarded as an emblem of hockey excellence to the best women's hockey team in Canada. However, in recent years competition in women's hockey has expanded rapidly in recent years and the American women's teams have been coming on strong, winning several world championships and Olympic gold medals.

As important as these milestones are and have been for the development of men's and women's hockey in North America, it is proving very difficult to establish a *professional* women's hockey league in North American due to the problem of creating a realistic pay scale for players so they can play hockey on a full time basis as most men do who are professionals. It has proven equally difficult to create professional hockey leagues in Europe for men and women since most European players prefer to come to North America to play for teams where the competition is keener and the salary levels are higher. Despite this, a professional men's league was eventually established in Europe when the Kontinental Hockey League (KHL) was formed in 2008. While it was established in Russia and consequently composed largely of Russian teams and players, it now has 29 teams from Russia, Eastern Europe, and Asia. They compete for the Gagarin Cup in much the same way that other teams and countries compete for the Stanley Cup, Spengler Cup, World Cup, Olympic gold medals, and other cups, medals, and awards.

In recent years, hockey has begun to expand and grow in other parts of the world, most notably in Asia, but also in the Middle East, Africa, and elsewhere. There is now an African Hockey Federation, an Arab Ice Hockey Federation established in 2008 with its headquarters in Abu Dhabi, and an Asia League Ice Hockey (ALIH) created in 2003 with nine teams from Japan, China, eastern Russia, and South Korea. With the Winter Olympics held in South Korea in 2018, it is clear that this will inevitably lead to a substantial increase in the growth of hockey in Asia in the years ahead. This may be especially true for China, which now has both men and women's professional hockey teams and has made a major commitment to expanding this sport throughout China that will undoubtedly mean sending effective hockey teams to compete with other countries in international tournaments and the Olympic Games in the years and decades ahead. For the first time in history, for example, two women's teams from China competed for the Clarkson Cup, which was given initially to Canadian women's teams only but more recently has expanded its scope to become more international rather than exclusively Canadian. This, in combination with the reduction in the number of fights and level of physicality in hockey and the increased emphasis on stick-handling abilities and passing and scoring plays has made hockey one of the most exciting sports to watch and play in the world.

Although the expansion of hockey internationally has been rather rapid in recent years, there are a number of reasons why getting to this point has taken such a long time and been so difficult. In the first place, hockey is and will probably always remain most popular in northern countries where climatic conditions and seasonal circumstances are far more congenial to this sport. In the second place, the game is much more expensive to play than many other competitive games because a great deal of costly equipment is necessary—skates, pants, sticks (which always seem to be breaking!), shoulder pads, helmets, masks, and so forth. Ice rinks are also expensive to construct, operate, and maintain.

The need to create a strong system for training, developing, and recruiting players also tends to limit international expansion. This system is made up of many different teams and leagues classified according to various age categories and other criteria, such as atom, pee-wee, bantam, midget, junior, senior, intercollegiate, industrial, and so forth. Some of these teams act as "farm clubs" for the professional teams. However, the fact that systems similar to those in North America and Europe are now being created in Asia, Africa, the Middle East, and elsewhere in the world augers well for the development of hockey on a worldwide basis in the future. This fact is not lost on Canadians, as the principal creators and primary contributors and promoters of this popular sport.

What is true for hockey is also true of another invention Canadians created and provided to the world that had its origins in hockey. It is "instant replay." This is a technology that enables the rebroadcasting of plays through recorded video shortly after they occur. This makes it possible for viewers to see plays a second, third, and fourth time moments after they happen, commentators to discuss plays by slowing down the frame count in order to allow more detailed analysis, coaches to go over plays with players both during and after games, and officials to examine and reverse calls that were made due to human error caused by the rapidity of events. Instant replay occurred for the first time in Canada, and can be traced back to 1955 when the Canadian Broadcasting Corporation and director George Retzlaff used a "hot processor" to develop footage for goals scored within thirty seconds so they could be replayed on *Hockey Night in Canada.*

Since that time, instant replay has evolved rapidly and is now used extensively throughout the world in virtually all sports as well as in many other fields. This makes it possible to get things right and reverse calls and mistakes by examining them immediately after they occur. Like hockey, the future of instant replay is assured, as is its rapid growth everywhere in the world.

Whereas hockey was created by many people over a very long period of time, basketball was created by one person and at a very specific point in time. That person was James Naismith, a Canadian who was born at Bennie's Corners near Almonte, Ontario in 1861.

Like many young men growing up in Ontario around that time, Naismith worked on farms during the summer and in logging camps during the winter. He was a keenly motivated and very determined person who eventually became a teacher, athlete, inventor, medical doctor, preacher, professor, and administrator. He was educated at McGill University and Presbyterian College in Montreal before being offered a job (which he accepted) to work in the United States as an instructor at the International Training School in Springfield, Massachusetts when he was in his early thirties.

While he was there, the director of the school asked Naismith to invent a game that students could play during the winter months and gave him two weeks to accomplish this feat. After a great deal of thought, Naismith invented basketball in 1891 and hand wrote 13 rules for this sport, the original copy of which was recently auctioned off by the Naismith International Basketball Foundation for $4.3 million to raise money for charity.

Basketball was an instant success and enjoyed by countless people almost immediately. While it was invented by a Canadian, it is Americans who have put their stamp on it and made it "their game." In fact, Americans have done for basketball what Canadians did for hockey, namely popularize this sport throughout North America and the world to the point where it is now deemed to be "America's game." Indeed, Americans created most of the rules, protocols, and procedures that govern this sport today.

Much of the expansion of basketball in the early years resulted from its rapid growth in colleges in the United States, the fact that it was played by both men and women (the first women's game was played in Smith College in 1893 for instance), and especially expansion of this sport in North America and other parts of the world through the YMCA. As far back as the late 1880s, the YMCA was involved in organizing a series of basketball exhibitions in France, Japan, India, China, and Persia (now Iran). In the decades to follow, the YMCA was actively involved in promoting basketball in its many missions throughout the world. Then, in 1936, basketball was officially recognized as an Olympic sport and played for the first time at the Olympic Games in Berlin, Germany. James Naismith was at these games. However, by this time he was working at the University of Kansas where he was coaching college teams and doing a great deal of pioneering work in physical education. In recognition of his outstanding contributions to the creation and development of basketball, Naismith was inducted into the Basketball Hall of Fame in Springfield, Mass. He also played a seminal role in the creation of the football helmet and many other devices that were designed to protect athletes and improve their performance.

Yet another major boost to the international expansion of basketball occurred in 1926 when the Harlem Globetrotters were created. As the name implies, their mission was to bring basketball to people in all parts of the world, largely through the opportunity this team provided for black players to demonstrate their remarkable skills and abilities at a time when they were barred from playing for virtually all teams in the United States except the ones they created themselves. Since 1926, the Globetrotters have entertained millions of people throughout the world with their amazing wizardry and artistry. They have played more than 26,000 exhibition games in 122 countries and territories around the world with no end in sight.

Spurred on by developments like this, basketball really started to take off after the Second World War. Among the key developments that have taken place since that time are the growth and development of the National Basketball Association (NBA) and the Women's National Basketball Association (WNBA), participation by the American "Dream Team" in the 1992 Olympics (something which did a great deal to promote interest in basketball in other parts of the world), marketing this sport actively in South America and Asia in general and Brazil, Argentina, China, the Philippines, and Australia in particular, creating leagues with outstanding players in South America, Europe, and other parts of the world, and televising games in all parts of the world and not just in United States, Canada, and Europe.

Due to developments like this, and many others, basketball is now the second-most played and watched sport in the world after soccer. It is played and watched by millions of boys and girls as well as men and women at every conceivable age and in every area of the world. Its popularity has been aided considerably by the fact that basketball is a very easy game to learn, understand, and watch, is not too expensive to play compared to many other sports, and has a relatively simple set of rules since it all boils down to "putting the ball in the basket" at both ends of the court. Basketball is especially popular in the United States, where millions of basketball courts and hoops are found in people's driveways and backyards from Maine to Florida and California to New York.

Around the world, there are countless amateur and professional leagues, as well as teams at every level of the educational spectrum. This is most evident in the United States at the college and university level. It is so popular at this level, in fact, that the entire month of March is devoted to celebrating what is called "March Madness." This involves competitions to determine the best men's and women's college and university basketball teams in the United States. Literally everyone gets caught up in the frenzy and excitement of this event, from the regional playoffs and the trip to the "sweet sixteen" to the "final four" and the national championship games for both men and women.

Like all sports, basketball has had its share of outstanding players and coaches who have done a remarkable amount to promote and popularize the game throughout the world. Included here are such legendary coaches as John Wooden, whose teams at UCLA in California won 10 NCAA (National Collegiate Athletic Assocation) championships in an incredible 12-year span, reeled off an 88-game winning streak, and achieved four 30-0 seasons, as well as Red Auerbach, who won nine NBA (National Basketball Association) championships as coach of the Boston Celtics over a ten year span and broke the colour barrier in basketball at the professional level by drafting the first ever African-American player to play in the NBA. But it is really phenomenal players over the years who have done the most to promote and popularize the game throughout the world—players such as Michael Jordan, Kareem Abdul-Jabbar, Wilt Chamberlain, Magic Johnson, Larry Bird, Bill Russell, Jerry West, George Mikan, LeBron James, and Stephen Curry to name a few. Their phenomenal feats on the court, as well as the branding as "superstars" that have resulted in huge sales of sweaters, shoes, T-shirts, shorts, and other memorabilia, have contributed considerably to making basketball one of the most played, popular, and watched sports in the world today.

While basketball has never taken off in Canada the way it

has in the United States, this has started to change and change dramatically in recent years. There are many reasons for this. Vince Carter's meteoric rise to fame as the slam-dunk champion when he played with the Toronto Raptors certainly helped to launch the grassroots revolution that has taken place in Canada in this sport. So did the remarkable career of Steve Nash, who was voted player of the year in the NBA on two separate occasions and was recently inducted into the Basketball Hall of Fame, as has the emergence of such Canadian stars as Andrew Wiggins, Nik Stauskas, Tylor Ennis, Kelly Olynyk, Jamal Murray, Cory Joseph, Trey Lyles, R.J. Barrett, Tammy Sutton-Brown, Natalie Achonwa, Kia Nurse, and many other Canadian men and women in the college and professional ranks in the United States. As Mike Hopkins, coach of the Washington Huskies men's team said recently, "Canada basketball is up-and-coming. They're competing against the U.S. for the best in the world." This certainly augers well for basketball in Canada and other parts of the world in the future.

Without doubt, the crowning achievement in the resurgence of interest in basketball in Canada was the Toronto Raptors winning the NBA championship in 2019, the first time a Canadian team (or any team outside the United States, for that matter) had ever won an NBA championship. After knocking off the Orlando Magic in round one of the playoffs, the Philadelphia 76ers in round two (with an unforgettable shot by Kawhi Leonard in the final second of the game that bounced around the rim of the basket four times before falling in), and the Milwaukee Bucks in round three, the Raptors faced the Golden State Warriors in the NBA finals. This team had won the NBA championship three times in a row after reaching the finals in each of the last five years.

And guess what? The Raptors knocked off the Warriors in six games to become NBA champions, with outstanding performances coming from Kawhi Leonard, Kyle Lowry, Pascal Siakam, Fred VanVleet, Serge Ibaka, Marc Gasol, and others, as well as Leonard winning the Most Valuable Player Award for the playoffs and Nick Nurse serving as head coach during the championship season after a number of years as assistant coach. It was an outstanding and courageous victory by a team that was committed to winning and simply never gave up. Not only was the whole country caught up in the excitement of this championship series run, but also it brought all of Canada together in a way that has seldom occurred before. And it was fitting. Not only was basketball invented by a Canadian but the first NBA basketball game was played in Toronto.

What is most remarkable about this is that all the activity that has taken place in the development of basketball throughout the world since it was invented in 1891 is the result of the creativity of one person—James Naismith—a Canadian who was born in a small town in rural Ontario named Almonte. This achievement is mindboggling in many ways. There is probably not a single person living anywhere in the world today (with the exception of very small children) who does not know about basketball and who has not watched or played the game at one time or another in their lives. How wonderful it would be if James Naismith could come back to watch a professional basketball game today to see how much the game has changed and been improved since he invented it more than a hundred and fifty years ago. And how much more wonderful it would be if Naismith could have been at the NBA finals in 2019 to watch the Torotno Raptors do Canada and Canadians proud by winning an NBA championship against all odds.

If Canadians have made many creative contributions to the development of hockey and actually did invent basketball, they also played a seminal role in the creation of lacrosse. Like ice hockey, lacrosse has deep roots in Canada, as well as a tradition that stretches back many centuries. It was originally created by Algonquin and Iroquois tribes along the St. Lawrence River and around the Great Lakes, although a similar game to lacrosse was probably played by Indigenous peoples much earlier.

Algonquin and Iroquois tribes called this sport *baggataway*, an Ojibwa word meaning "ball." When it is played outdoors it is called "field lacrosse," and when it is played indoors it is called "box lacrosse," or "boxla." The rules for the game were set out in a formal sense in 1860 by George Beers—Canada's "Mr. Lacrosse"—who organized a lacrosse convention in 1867 that led to the creation of the National Lacrosse Association. He also wrote a book—*Lacrosse: The National Game of Canada*—in 1869 that generated a great deal of interest in this sport. By the end of the nineteenth century, lacrosse had become Canada's unofficial "national game" and was played in many parts of the country. This led to the creation of the Minto Cup in 1901, yet another sporting trophy named after a Governor-General— Lord Minto in this case. The Minto Cup was awarded to the senior amateur lacrosse champion in Canada for many years, and is now awarded to the junior champion. This was followed, in 1910, by the creation of the Mann Cup, which is now awarded to the senior amateur lacrosse champion.

Despite the flurry of activity at the end of the nineteenth and beginning of the twentieth century, lacrosse has never gained traction in Canada or the world the way ice hockey and basketball have. Nevertheless, it did achieve some popularity in Ontario and British Columbia, where intense rivalries evolved between particular teams and specific leagues. As a result, teams like the Peterborough Lakers, Oshawa Green Gaels, Burnaby Lakers, Victoria Shamrocks, Brampton Excelsiors, Six Nations Chiefs, and the New Westminster Salmonbellies dominated the Minto and Mann cups for many years. Then, in the 1960s and '70s, attempts were made to create a professional lacrosse league in Canada. After several attempts failed, the North American Lacrosse League was finally formed in the 1990s and was dominated for many years by the Toronto Rock.

It was probably for reasons such as these that in 1994 Parliament declared lacrosse to be Canada's "national *summer* sport." This immediately caused a great deal of confusion and controversy throughout the country because no one was quite sure whether lacrosse or hockey is really Canada's "national game." Officially speaking, it is lacrosse in the summer and hockey in the winter. However, unofficially, everyone knows that hockey is really Canada's "national game," regardless of the season of the year or the progress that has been made in the development of lacrosse in Canada, United States, and other parts of the world in recent years, particularly at the youth, college, university, and professional levels.

One of the major reasons for lacrosse's growing popularity is the fact that it is now played by women as well as men and doesn't cost a great deal to play. It is growing rapidly in popularity in the United States where it has become the American "sport of choice" for young people according to the *Wall Street Journal*, which recently noted there are more than three-quarters of a million children, teenagers, and students at the NCAA or college level playing this game. Many colleges and universities have men's and women's lacrosse teams, and the game is rapidly gaining popularity in other parts of the world as well, especially in Europe, Russia, China, and more than 60 other countries.

Then there is curling. While Canadians didn't invent this popular game, they have made many highly creative contributions to its development over the years. Whereas some people claim curling was created in Belgium or Holland and point to paintings by Pieter Breugel the Elder as proof of this, most believe it was invented in Scotland where it has been played, enjoyed, and given special status for several centuries. Nevertheless, the game has been played in Canada for more than two centuries, and even the Scots would have to admit that Canadians have made many of the most important and innovative contributions to the development of this sport over this period of time.

Canada's long and distinguished tradition in curling goes back to the beginning of the nineteenth century. Indeed, the first sports club in Canada was a curling club—the Montreal Curling Club, which was established in 1807 by Scots who came to Canada and were involved in the fur trade (most likely the North West Company), thereby explaining how curling made its way from Scotland to Canada. By 1840, numerous curling clubs had been created in what was soon to become Canada and the first book on curling—*The Canadian Manual on Curling*—was written by James Bicket.

Interprovincial curling matches—the main feature of the game played in Canada today—began in 1856. This was followed by the first ever international curling bonspiel. It was held in 1865 and played between teams from Canada and the United States in Buffalo, New York. While curling was originally played outdoors, it moved indoors around 1850. By this time, Canada was considered by many to be the "Eldorado of Curling." The first women's curling club was founded in Montreal in 1894, and the

W. D. Macdonald Tobacco Company hosted the first Dominion Championship by awarding the Macdonald Brier Tankard to the winner in 1927. This was followed by the establishment of the Dominion Curling Association in 1935, which was renamed the Canadian Curling Association in 1968. However, the most significant event of all as far as the international development of curling is concerned was the championship game played between Canada and Scotland in 1959. This set the stage for the creation of world curling championships that are incredibly popular among curlers, curling enthusiasts, and the public at large in many parts of the world today.

Just as other sports have their heroes and superstars, so does curling. Some of the best known curlers in the Canadian case are Ernie Richardson and Sandra Schmirler from Saskatchewan, Ed Werenich and Jeff Stoughton from Manitoba, Dan Holowaychuk and Kevin Martin from Alberta, Brad Gushue from Newfoundland and Labrador, and Glen Howard from Ontario.

Ernie Richardson and his team won four Canadian championships in five years, as well as four world championships in the late 1950s and early 1960s. Richardson is often described as "the best curler Canada has ever produced and the world has ever seen." This might be contested by people in Ontario who point out that Glenn Howard from Tiny, Ontario won four world championships, four Briers, and 15 Ontario provincial championships, including eight straight championships between 2006 and 2015. It might also be contested by followers of Sandra Schmirler, whose team won three world and Canadian titles, as well as a gold medal at the Olympic Games in Nagano, Japan in 1998. Schmirler was often described as "the queen of curling" before her untimely death in 2003 that is still lamented by many people in the curling community today.

And there is more. Much more. Ed "the Wrench" Werenich and his team won one Brier and two world championships, and is generally regarded as the man who could perform most effectively in competitions by "turning up the heat" when it was necessary to do so. Not to be outdone, Brad Gushue and his team won the gold medal at the Olympic Games in Turino, Italy in 2006, thereby proving that it is not only curlers from western Canada or Ontario that are capable of winning Olympic championships. And in 2010, former World Champion Kevin Martin and his team from Alberta won the gold medal for curling at the Olympic Games in Vancouver, thereby erasing the painful memory of a heartbreaking loss to Norway at the Olympics in 2002 when they had to settle for a silver medal. Interestingly, Canada's men curlers won the World Championship in curling in 2010 and again in 2011, when Jeff Stoughton won his second world championship by beating a tough rink from Scotland, thereby establishing Canada's international supremacy in this sport. This supremacy was maintained by Jennifer Jones. She has won six national championships, won the World Championship in 2008, and was undefeated as the skip of the Canadian women's team in winning the gold medal at the Olympics Games in Sochi in 2014, much as Kevin Martin did as skip of the men's team at the Olympics in 2010. In all, Canadian men and women have won more than 45 gold medals at world curling championships, far more than any other country.

Canada can also lay claim to having "the most effective and creative" icemaker in the curling world—Clarence "Shorty" Jenkins, who died in 2013 and was inducted into the World Curling Hall of Fame in 2018. He is known as the "Wizard of Ice," "wore a pink Stetson, slept in many rinks, and changed the face of curling rinks throughout the world," according to Russ Howard, a two-time Canadian and world champion.[12] He accomplished this feat by knowing the specific rolls, features, conditions, and details of most curling rinks and ice surfaces throughout the world far better than anyone else.

According to Doug Clark, author of *The Roaring Game: A Sweeping Sage of Curling*, Jenkins did more to revolutionize the game of curling than anyone else in the world. This resulted from Jenkins' disappointment at not making the Ontario men's championship in 1974 because of what he felt was "bad ice." Here is what Clark describes as the consequences of this disappointment:

> Shorty left that qualifier determined to find a better way to make ice. In short order, he revolutionize curling with a stop-watch and his powers of observation and common sense. He concluded that "a curling rock is smarter than a human being," and he shared that epiphany in ice-making courses he teaches around the world. While good skips can unerringly "read" the ice, rocks "feel" it, reacting immediately to any change, no matter how great or small. That may not sound like rocket science, but before Shorty, no one had discovered or articulated that basic fact of physics and common sense. No superlative adequately describes the global impact on the roaring game of that epiphany by the man exalted by curling icon Doug Maxwell in the *Canadian Curling News* as "the King of Swing."[13]

---

12. Kenneth Kidd, "Chilling out with the Wizard of Ice," *Toronto Star*, November 8, 2009 (Section IN).
13. Doug Clark, *The Roaring Game: A Sweeping Saga of Curling* (Toronto: Key Porter Books, 2007), pp. 127–128.

Clark goes on to explain in great detail Jenkins' incredible international impact on the game:

> As the man most responsible for the evolution of curling from the Stone Age to the Ice Age, Jenkins became visible early for his flamboyant, vividly-hued curling outfits, topped off with his trademark Stetson.
>
> Honoured as a curling pioneer, inducted into the Canadian Curling Hall of Fame, he has been further enshrined in pop culture, on television, and the big screen: he was satirized on the *Canadian Air Farce*, played himself in a national ad plugging Tim Horton's coffee, and made a cameo appearance in *Men with Brooms* [one of Canada's largest grossing movies, starring the outstanding Canadian actor Leslie Nielsen]. In the process, he breathed fresh air into a traditionally staid sport and, without stealing the thunder of the on-ice stars of the roaring game, has perhaps transformed the sport further, and faster, than any other.[14]

It is contributions like these that help to explain why Canada and Canadians have had such an incredible impact on the development of curling in most parts of the world. Not only have many Canadian curlers—such as Doug and Gerry Graham, Ian MacAulay, Kirk Smyth, Keith Wendorf, and especially the Merklinger clan—travelled to other countries and parts of the world to impart their skills and knowledge of the game to help others create and play it, but they have also been responsible for the rapid growth of curling and the improvement of ice surfaces from New Zealand and Australia to Russia and the Scandinavian

---

14. Doug Clark, *The Roaring Game*, p. 128 (insertion mine).

countries. As a result, it is now estimated that there are more than 1.2 million curlers in the world. While an amazing 90 percent of these are registered in Canada with more than 1,100 clubs, the United States now boasts as many as 13,000 curlers and the numbers keep growing rapidly in other parts of the world as well. Although it is a sport like ice hockey that is played far more in northern countries, it is rapidly gaining acceptance and participation in other parts of the world.

As a result of the growth of curling throughout the world, Canada's supremacy in men's and women's curling at the Olympics and world championships has been increasingly challenged by curling teams from other countries in the world in recent years. Not only have outstanding Canadian curlers been going to other countries and parts of the world to teach curling for many years now but also Canadians have also coached teams from other countries at most major curling events. For example, at the Olympic Games in South Korea in 2018, six of the top seven women's teams in the world were coached by Canadian curlers, including the British team by Glenn Howard, the Japanese team by J. D. Lind, the American team by Al Hackner, and the South Korean team by Peter Gallant. While this might be disheartening to Canadian curlers, it is reassuring to know that the number of curlers and quality of curling around the world have been growing steadily as a result of Canada's historical and contemporary creative ingenuity and expertise in this sport and willingness to share this with other curlers, teams, and countries in the world.

Skating is another sport where Canadians have excelled and made many creative contributions to the world over the years. This is not only true for figure skating, but also for speed skating.

Canada's most important contributions to the world in figure skating have been in the development of numerous skaters who have inspired countless skaters in other countries and parts of the world by winning numerous world championships and gold, silver, and bronze medals at the Olympic Games, landing many very difficult jumps for the first time in international competitions, changing the way figure skating is choreographed, and the creation and development of Skate Canada, a remarkable model for training men and women in single skating, pairs staking, ice dancing, and synchronized skating in different parts of the world.

Canada's long tradition in figure skating began in the late nineteenth century when Louis Rubinstein, who was the first real figure skating hero in Canada and the star of the Victoria Rink's figure skating club in Montreal, was amateur champion in the United States in 1885 and 1889, as well as unofficial world champion in 1890 when the first world championships were held in St. Petersburg, Russia. Half a century later, Barbara Ann Scott became Canada's and the world's most-admired and cherished figure skater when she won the European, Olympic, and world championships within a period of six weeks in 1948, a feat unequalled by any other skater.

This set the stage of many Canadian Olympic and international figure skating champions who followed in her footsteps, such as Barbara Wagner and Robert Paul, Otto and Maria Jelinek, Donald Jackson, Karen Magnussen, Toller Cranston, Kurt Browning, Elvis Stojko, Jamie Salé, David Pelletier, Patrick Chan, Scott Moir and Tessa Virtue, and many others. Not only did Moir and Virtue win the gold medal at the Winter Olympics as well as the World Championships in 2010, but also they won the gold medal at their last Olympic performance in Pyeonychang, South Korea in 2018, as well as winning many gold, silver, and bronze medals and world championships between those years. They are the most decorated figure skaters in Olympic history, thereby providing motivation and inspiration to myriads of male and female figure skaters from Canada and other countries in the world.

And this isn't all. Canadian skaters have also demonstrated a great deal of creativity, imagination, and skill by landing some of the most difficult jumps in figure skating competitions over the years. Included here are Maria and Otto Jelinek, the first pair to perform lifts with several rotations; Donald Jackson, the first skater to successfully land a triple lutz in international competition (a jump from a backward outside edge, assisted by the toe pick, with three revolutions in the air); Petra Burka, the first woman to execute a triple salchow in competition (a jump from the back inside edge of one foot to the back outside edge of the other foot with three rotations in the air); Vern Taylor, the first person to successfully land a tripe axel in competition; Brain Osler, winner of the world title in 1987 and the first competitor to complete two triple axels successfully in a single program; and Kurt Browning, who executed the first quadruple jump in competition in 1988.

Among this illustrious group of world-renowned figure skaters, Toller Cranston is generally regarded by many as the most creative and innovative of all, although arguably they all have been very creative in their skating, programming, and jumps in one form or another. However, as a well-known visual artist as well as an outstanding figure skater, Cranston transformed figure skating in the world when he skated at the Olympic Games and world championships, largely by changing the choreographic approach to the "long program" and influencing countless skaters throughout the world who followed in his footsteps. Many of his creative achievements in this area are set out in the book, *Toller Cranston, Zero Tollerance: An Intimate Memoir by the Man Who Revolutionized Figure Skating*.

Canadian figure skating successes like this, and many others, would not have been possible without the creation of an outstanding organization for training figure skaters. This is Skate Canada, the largest association of its type in the world. From two clubs in 1914, it grew to a membership of 360 clubs in 1967 and 1,410 in 1986. By 2017–18, this had catapulted to 179,023 members, 5,758 accredited coaches, and 1,162 clubs. This valuable organization offers programs for skaters of all ages and at all levels, including beginners as well as power, synchronized, and advanced skaters.

It is not only figure skating where Canada has recorded many international successes, has a long and distinguished creative tradition, and influenced skaters in other countries in the world. It is also true for speed skating. The first Canadian speed skater to win a world championship was Jack McCulloch of Manitoba. He won the Canadian speed skating championship in 1893, the U.S. Nationals in 1896, and the world championship in Montreal in 1897. Other Canadian speed skaters to achieve international success were Charles Gorman from St. John, Fred Robson from Toronto, who held ten world speed skating records between 1899 and 1916, and Frank Stack from Winnipeg who was a world record holder in the early 1930s.

This remarkable tradition was carried on in excellent fashion and advanced considerably decades later. Gaëtan Boucher from Ste.-Foy, Quebec won a silver medal at the Olympic Games in Lake Placid, New York in 1980, as well as a bronze and two gold medals at the Olympic Games in Sarajevo, Yugoslavia in 1984. Catriona Le May Doan won numerous gold, silver, and bronze medals at the Olympic Games and other international competitions. As remarkable as these feats were, they were topped by Cindy Klassen—yet another famous speed skater from Winnipeg and Manitoba!—who won five medals—a gold, two silver, and two bronze—at the Olympic Games in Turin, Italy in 2006. She also won five medals and the overall title at the world championships in the same year, establishing two world records in the 1,000 and 3,000 metres as well as at the World Cup with the overall title in the 3,000/5,000 metres and second in the 1,500 metres. In recent years, Canadians have also been winning many team events in speed skating, especially Isabelle Weidemann, Ivanie Blondin, and Keri Morrison who won a recent World Cup in this event.

It is feats like these that have contributed significantly to the growing interest in speed skating in Canada and many other parts of the world. Thanks in no small part to the impetus provided by Canada and Canadian speed skaters as well as those of other northern countries, this sport is now very popular in virtually all parts of the world, much like figure skating. As a result, many speed skaters and figure skaters come from other countries to practice and compete in Canada rather than remaining in their own countries due to Canada's superb facilities, training programs, teachers, coaches, and other opportunities in these two closely related sports.

It is not surprising given Canada's mountainous areas, snowfalls, and long winters that skiing is another sport where Canadians have excelled and been very creative over the years in inspiring others. In 1958, Lucile Wheeler of Sainte-Jovite, Quebec won gold in both the world giant slalom and downhill in Bad Gastein, Austria. Two years later, Anne Heggtveit, whose father Halvor was a champion cross-country skier, won the slalom at the Winter Olympics in Squaw Valley, California in 1960. She was also the world title-holder in the slalom and alpine combined. This paved the way for a number of outstanding skiers who came somewhat later, most notably Nancy Greene, who was the first Canadian World Cup winner in 1967 and won gold and silver medals at the Olympics in 1968, and the "Crazy Canucks"—Ken Read, Steve Podborski, Dave Irwin, Jim Hunter, and Dave Murray—who earned more than a hundred Top 10 World Cup finishes between 1978 and 1984, 39 of them in the top three. Also included in this group are Kerrin Anne Lee Gartner, who won gold at the 1992 Olympics, and Chandra Crawford who won gold at the Olympics in 2006. These achievements were sustained and enhanced at the 2014 Olympics in Sochi, Russia, where the three Dufour-Lapointe sisters—Justine, Chloé, and Maxime—won gold and silver medals as well as a twelfth place finish respectively, and Erik Guay won the World Alpine Skiing championships at St. Moritz, Switzerland in 2017.

In recent years, Canadian skiers have been performing extremely well and making creative contributions to some of the newer skiing events, such as giant slaloms, bigair, slopestyle, halfpipe, parallel, and speed snowboarding for men and women. The most outstanding Canadian skier of all time, however, is undoubtedly Mikaël Kingsbury. Not only did he win a gold medal in moguls at the Pyeongchang Olympics in 2018, but he is the first man to have won medals in both moguls and dual moguls at the world championships in 2015 and 2017, as well as two gold medals in these two events at the world championships in Deer Valley, Utah, in 2019. He has also won the most medals at the freestyle world championships of any male competitor in history, having won a medal in seven of eight events, but also seven consecutive World Cup titles and the most World Cup medals ever, with more than 75 medals overall and more than 50 of them golds. However, the crowning achievement in his life to date may be winning the Lou Marsh Award at the end of 2018 as Canada's "athlete of the year." This award goes back to 1936 and is chosen by a panel of sports journalists and experts from all parts of the country.

Not surprisingly, Canadians have also excelled at bobsledding, where Pierre Lueders and David Bisset paved the way and Kaillie Humphries and Heather Moyse won gold medals at the 2010 and 2014 Olympics. In 2015 Humphries became the first woman to compete in international competitions against men in a bobsled race, and later in the same year, won the prestigious Lou Marsh Award as Canada's best athlete.

According to many experts, Canadians also played a pivotal and highly inventive role in the creation and development of synchronized swimming throughout the world. This took place initially in the 1920s when a group of Canadian women developed what they called "fancy swimming" when they were preparing for their Royal Life Saving Society diplomas. It was at this time that the first competition in this sport occurred in Montreal in 1924. Then, in 1938, Margaret (Peg) Sellers, a water-polo player and diver, devised a standard set of rules for this graceful and exquisite sport. In the 1940s, the term "fancy swimming" was changed to "synchronized swimming." It debuted at the Olympics in 1952, and became an official addition to the Olympic Games in 1984. The undisputed master of this sport has been Carolyn Waldo of Canada. She won a silver medal in synchronized swimming at the Olympic Games in 1984 and then two gold medals at the Olympic Games in 1988 in this popular and rapidly growing sport.

Another sport where Canadians have performed very well

and made creative contributions with global implications and consequences are the Paralympic Games, particularly the Winter Paralympic Games. The growth of these games has been phenomenal. Prior to 1988, they were played in very modest venues as well as in different cities from the Winter Olympics, with few spectators and little coverage and promotion. This changed in 1988 when these Games were held in Seoul, South Korea. While there was still very little coverage and promotion, it was the first time the Winter Olympic Games and the Winter Paralympic Games were held in the same place and at the same time, thereby expanding the audiences for the latter Games as well as their exposure and popularity. Since that time, remarkable growth has taken place in the Paralympic Games. In 2018, the Winter Olympic Games and Paralympic Games returned to South Korea—this time in Pyeongchang, where audiences were large and promotion was exceptional. This led to joint agreements between the International Olympic Committee (IOC) and the International Paralympic Committee (IPC) to engage in close cooperation and collaboration on all developments in this area in the future.

 The performances of Canada's Paralympic athletes at both the 2010 Paralympics in Vancouver and the 2018 Paralympics in Pyeonychang were remarkable. They performed brilliantly at the Vancouver Paralympics, winning 19 medals including three gold medals in cross-country skiing by Brian McKeever, a blind skier who was guided by his brother; a gold medal in curling by Jim Armstrong and his team; and an incredible five gold medals by Lauren Woolstencroft, who was born with both legs missing below the knees and her left arm missing below the elbow. These Games were also marked by many other outstanding and creative achievements, as well as some timely recognition for Terry Fox and Rick Hansen, both discussed earlier in the book. But the greatest achievement of all during the Paralympics in Vancouver was undoubtedly the breakthrough that occurred in the size of the audiences at these Games, as well as the commitment that was made to recognizing Paralympic competitors as "splendid athletes" rather than "disabled people."

This success was expanded even more at the 2018 Games in South Korea where Canada's Paralympic athletes won a total of 28 medals, including six by Mark Arendz, who won a medal in every event he entered and was Canada's flag bearer at the closing ceremonies, as well as Brian McKeever, once again, who won three gold medals and one bronze medal and was Canada's flag bearer at the opening ceremonies. McKeever has won 17 medals, including 13 gold medals—the most ever won by a Winter Games Paralympic athlete. His performance and the performances of other Paralympic athletics from Canada leave no doubt that Canada's Paralympic athletes have had a powerful impact on people and countries in other parts of the world and the world as a whole.

While many Canadians enjoy sports that are competitive and are actively involved in playing and watching them in person and/or on television on a regular basis, others are involved in other types of recreational activity, such as gardening, camping, hiking, boating, cottage life, and travelling to different parts of Canada and the world. This explains why Canadians probably own more recreational equipment and facilities on a per capita basis—cottages, cabins, Coleman stoves, tents, canoes, kayaks, motor boats, Ski-Doos, skis, snowshoes, snowboards, toboggans, and the like—than any other people in the world. This is not surprising in view of the pristine nature of much of the country's geography, as well as the countless rivers, lakes, mountains, forests, and wilderness areas that exist across the country.

Nowhere has this recreational activity been more in evidence in an historical sense than in the Laurentian region of Quebec. Many well-known Quebecers have contributed to the creation of recreational activities, trails, and facilities in general and winter recreational activities, trails, and facilities in particular. Undoubtedly, the most famous of these was Herman Smith "Jack Rabbit" Johannsen. Born in Norway, Johannsen was a superb skier who came to Canada at an early age and settled in Montreal. He created numerous trails for alpine and cross-country skiing in the Laurentian area and organized countless recreational events and festivities there. Living more than a hundred years, he is recognized as one of the truly great creative pioneers of winter recreation in Canada. To this list, however, should be added Joe Ryan and Mont-Tremblant, one of eastern North America's most outstanding ski resorts; Emile Couchand, who created the continent's first mechanical ski lift; and Curé Labelle, a priest who created the P'tit train du Nord that carried skiers on special trains between 1920 and 1940 to popular skiing destinations in Quebec, as well as paved the way for the creation of the longest linear park in North America with 42 kilometres of cross-country ski trails and 114 kilometres of snowmobile trails.

In addition to this, incredible creative energy and effort has gone into planning and marketing the many different recreational activities, events, and facilities that exist in Canada. Visits to art galleries, museums, and historic sites are very popular. Consequently, most towns, cities, and provinces have well-developed recreational and tourist facilities, organizations, programs, and amenities.

Canada also has one of the largest, most extensive, and best organized systems of parks, conservation areas, and historic sites in the world. This is due largely to visionary and creative achievements on the part of officials at Parks Canada, as well as in the federal, provincial, territorial, and municipal governments. In fact, Canada was the first country in the world to have a *national park service*. At the centre of this service is Parks Canada, which was established in 1911 as the Dominion Parks Branch in the Department of the Interior. It was fortunate to have had James B. Harkin, known at the time as "the father of national parks," as its first commissioner. He was a Canadian-born journalist who became a civil servant and served from 1911 to 1936, where he was responsible for the creation of a number of national parks across the country. While his life was filled with many achievements, it was also marred by certain controversies, especially when Wood Buffalo National Park was established in 1922 as the second-largest park in the world. It sparked a major outrage and debate among the Indigenous peoples of the country that still persists and resonates strongly today, thereby reinforcing the need for full consultation with Indigenous peoples on matters related to their interests and cultures in general and their affection for and connection to the natural environment and the land in particular.

Canada also has two of the oldest national parks in the world—the Cave and Basin Hot Springs created in 1885 which provided the basis for the establishment of Banff National Park in 1887, and Yoho National Park, created in 1896—both of which were established before the country's national parks service existed. Since that time, the service has grown rapidly. As of 2018, Canada had 47 national parks—composed of 40 actual national parks and 7 national park reserves—and 10 new parks and five new marine areas either in the active planning stages or actually completed in such places as Wager Bay, Northern Bathurst Island, Great Slave Lake, the Northwest Territories, and Nunavut. Consequently, visits to national, provincial, and municipal parks, as well as conservation and marine areas, are extremely popular with residents and tourists alike. So are visits to cultural monuments and historic sites, most notably UNESCO World Heritage sites such as Quebec City, Lunenburg, L'Anse aux Meadows, Nahanni National Park, Dinosaur Provincial Park, Sgaany Gwaii, Head-Smashed-In Buffalo Jump, the Canadian Rocky Mountain Parks, and Gross Morne National Park.

The crowning achievement in this service is undoubtedly the recent completion of the Trans Canada Trail, or TCT, which at more than 20,000 kilometres is the longest recreational trail in the world. Almost 30 million Canadians—roughly 80 percent of the country's total population—are able to access this trail within 30 minutes or less. The trail connects many of Canada's diverse habitats, exquisite landscapes, unique communities, and remarkable natural treasures together.

As we have seen, Canada and Canadians have made many creative contributions to the world in sports and recreation. In some cases, it is specific sports that have had the greatest international impact, especially hockey, basketball, lacrosse, curling, figure skating, speed skating, skiing, bobsledding, and synchronized swimming. In other cases, it is specific recreational activities that have had the greatest world impact, such as the creation of the national park service and the transnational trail which will probably act as a model for other countries in the future as they develop similar facilities and capabilities of their own. With the rapid growth of sports and recreation in all parts of the world, it is likely that Canada will continue to play a prominent, creative, and exemplary role in this area in the years and decades ahead.

Chapter Eight
Education, Politics, and Political Milestones

Building effective educational and political systems in Canada has been a long and arduous process. Not only has it been necessary to overcome many difficult transportation, communications, and geographical challenges, but it has also been imperative to address some very contentious religious, linguistic, ethnic, racial, and demographic issues.

In the process of confronting these challenges and issues, Canadians have exhibited a great deal of creativity in education and politics—not necessarily in the general sense, as Canadians have tended to follow the lead and duplicate the models in existence in Great Britain, France, and the United States, but certainly in the specific sense, as a result of the remarkable amount of creativity that has been needed in these two areas to come to grips with Canada's colossal size and diverse population.

In education, this has been especially true of the creativity needed to deal with the requirements of different religious, ethnic, racial, cultural, linguistic, and age groups as well as the development of distance, correspondence, adult, and early childhood education, helping people with special needs and major disabilities, and, most recently, education related to artificial intelligence and "deep learning."

Religion and ethnicity played a powerful role in the development of education in Canada, particularly in earlier times but also to a certain extent even today. Following the amalgamation of Lower and Upper Canada into a single province in 1841 but before Canada was created as a country in 1867, for instance, efforts were made to bring together the major religious and linguistic elements of the population at that time—Catholics and Protestants as well as French and English—in the realization of a single, integrated educational system. It was a noble effort, but a resounding failure in the end. As a result, the Common School Act specified that members of a religious minority in any local school area or district had the right to establish their own school boards, open their own schools, and receive provincial government grants for this purpose. This became the legal basis for the dual system of Protestant and Catholic schools, boards, and commissions that developed in Canada East and persisted in Quebec after Confederation. It also served as the justification for the development of "separate schools" in Canada West (later Ontario). While many Canadians think this guarantee of educational autonomy for a religious minority was a concession made to Catholics—primarily in Quebec where most of the population was Roman Catholic—it was actually a concession made to the Protestant minority in Montreal who wanted to establish their own schools and have this right incorporated into the Common School Act.

This type of bifurcated educational system has been maintained in different parts of the country ever since, despite criticisms leveled against it from time to time. By the time of Confederation in 1867, the pattern for educational development in Canada had been set, especially with respect to constitutional authority over education and its relationship to religion, religious institutions, and minority groups. Constitutionally, education was to be a provincial rather than federal responsibility. Catholics and Protestants were guaranteed denominational schools supported by provincial governments where they constituted a minority and numbers warranted it. If any province refused to recognize this guarantee, the British North America Act made it possible for the federal government to intervene and enforce this requirement. This provision profoundly affected the development of educational institutions and systems in the west of Canada, especially Manitoba, Saskatchewan, and Alberta, where the question of separate schools or a single integrated school system was played out in earnest.

Meanwhile, another problem was percolating on the educational front that required a creative response. It was the problem of preserving the languages and heritages of ethnic minorities and communities other than French and English. Many of the immigrants who came to Canada shortly after Confederation were from European countries other than England and France. Like the English and French, they wanted their languages and heritages preserved and protected and the country's emerging elementary and secondary schools became the focus of their attention and concern. In the west, strong pressures were exerted to teach Polish and Ukrainian in schools in areas where these ethnic groups were concentrated. In Toronto and Montreal, similar efforts were made to teach Hebrew and preserve Jewish customs, traditions, and culture where numbers warranted it.

As these developments indicate, multiculturalism is not a new phenomenon in Canada. It has deep roots in the country's history, culture, geography, and demography, and is manifested in the many multicultural activities that occur in most of the country's schools across the country at present, from activities concerned with Black history month, the Holocaust, the Indigenous peoples, and minority cultures to specific courses, programs, and projects designed to celebrate different ethnic groups and their distinctive customs, languages, traditions, and so forth.

What is true for religious, heritage, language, and multicultural education is also true for distance and correspondence education, the use of audio-visual and digital technologies in education, teaching people with special needs and disabilities, and adult and early childhood education. A few examples should suffice to illustrate the nature and extent of this particular dimension of education in Canada and its implications for education in other parts of the world, especially parts of the world where similar demographic and geographical conditions and characteristics prevail.

For many years, railroad cars were used in Northern Ontario to educate people in isolated areas. These cars were outfitted with classrooms, shunted onto sidings, and remained in remote areas and communities for weeks at a time, much like "museum-mobiles" did a century later. Moreover, a number of universities distinguished themselves in distance and correspondence education and became international leaders in this field, most notably Athabasca University, an open university which instructs by means of self-study; Ryerson University, which ranks high in terms of the number of distance students; and the Université du Québec and University of Waterloo, which have well-established distance education programs. The latter institution has become well-known and emulated throughout the world for this, as well as for its very successful cooperative programs. These programs combine academic training in the classroom and practical experience on the job. They also utilize many audio-visual resources and computer and digital technologies.

The Toronto District School Board is also renowned for

its use of audio-visual equipment and computer and digital technologies, in large part because Toronto was one of the first cities in the world to make coaxial cable installations available to households that required them. This made it possible to deliver specially designed television programs by cable companies to any classroom in the city, as well as to homes where children were not able to attend school. Toronto was also the first city in Canada—and the second in North America—to make kindergarten an integral part of the public education system. This occurred back in 1883. Two years later, Ontario became the first province or state in North America to approve the offering of kindergarten programs in all elementary schools. Toronto was also one of the first municipalities in the world to establish a permanent program for convalescing students in collaboration with the Hospital for Sick Children.

To this list of creative responses to educational needs and circumstances should be added Mount Allison University in New Brunswick, which was the first university in Canada to combat gender discrimination in education by granting a degree to a woman, namely Grace Annie Lockhart in 1875. Canada also played a seminal role in the development of education for people with special needs and physical disabilities. In 1857, for instance, Halifax created the first School for the Deaf, and in 1871, two schools were built for the blind, one in Fredericton, New Brunswick and one in Brantford, Ontario.

While these developments have brought Canada a certain amount of recognition in education in other parts of the world, the country is especially well-known for its highly creative contributions to adult education. This can be traced back to the latter part of the nineteenth century when extramural and extension courses were offered by Queen's University in Ontario, night classes were provided by the YMCA, many farmers' institutes were established with educational and training mandates, and Adelaide Hoodless founded the first Women's Institute in Stoney Creek, Ontario. Predicated on the goal of helping women to obtain formal education while simultaneously carrying out their household responsibilities, the Women's Institute organized women in rural areas across the country and enabled them to speak with a united voice with respect to educational matters and reforms. Another important contribution in this area was made in 1899 when Frontier College was created. It was founded by Alfred Fitzpatrick, a Presbyterian minister working with Queen's University in Kingston, Ontario, and was one of the first colleges to concern itself with adult education in remote areas.

An even more valuable contribution to adult education was made when Dr. Roby Kidd founded and developed the Canadian Association for Adult Education. Much more important in an international sense, however, was the leadership he provided by working closely with authorities at UNESCO and other educational institutions in the development of adult education in other parts of the world and the world as a whole as indicated earlier. It is largely as a result of Kidd's pioneering efforts and global achievements in this area that the International Council for Adult Education is situated in Canada and has its headquarters in Toronto.

A comparable contribution was made to early childhood education by Fraser Mustard. In the late 1990s, Mustard co-authored a seminal report with Margaret McCain for the Government of Ontario called *The Early Years Study* that emphasized the promotion of early childhood development. This led to the creation of a province-wide full-day kindergarten program—previously it was usually half-day or every other day—as well as providing public funding for the development and education of children at a very early age. Mustard was also involved with governments in Australia, Mexico, Brazil, Chile, and China, as well as the World Bank, the Inter-American Development Bank, UNICEF, and the Aga Khan University in Pakistan, in promoting the crucial importance of early childhood education and its development on a worldwide basis. Jim Grieve, a former public school board director of education in charge of implementing full-day kindergarten throughout Ontario, called Fraser Mustard "Canada's gift to the children of the world." He went on to say that there is not a country he has visited in any part of the world that does not know about Fraser Mustard and his remarkable accomplishments in this very important dimension of education.

Interestingly, Canada and Canadians are playing a highly creative, proactive, and pioneering global role in the educational and intellectual development of artificial intelligence and deep learning and have been doing this for some time now. Intelligence and learning manifested by machines, for instance, can include such examples as computers playing chess and checkers with human beings and winning, as well as the IBM's supercomputer named "Watson" that defeated two of the greatest *Jeopardy* champions in the world by a wide margin several years ago.

Artificial intelligence is a rapidly expanding and highly complex field that includes computational creativity, machine and deep learning, neutral networks, fuzzy systems, evolutionary algorithms, Bayesian networks, chaos theory, robotics, and a great deal else. Many think artificial intelligence and deep learning will play a crucial role in the world of the future since they have countless theoretical and practical applications ranging from scientific and medical advances as well as speech and language recognition and retention to economic, social, educational, and political planning, policy, and decision-making. However, many authorities are equally concerned about the very real threat posed by artificial intelligence, believing that intelligent machines may eventually escalate out of control, escape human management, have lives of their own, and produce much more unemployment in the years and decades ahead.

While the United States and China are deemed to be the superpowers in the rapidly expanding fields of artificial intelligence and deep learning in a quantitative sense, Canada is an international hotbed for developments in this area and is considered by many to be a creative leader in this field, as well as in computer algorithm-powered technologies. As Richard Zemel, a senior fellow at the Canadian Institute for Advanced Research said—the Institute's first president and co-founder was, incidentally, the same Fraser Mustard discussed earlier— "Canada is home to some of the brightest minds in the field of artificial intelligence and has been in the academic forefront of the field of AI and deep learning for over 30 years." He went on to say that in the areas of machine, reinforcement, and deep learning, Canadian researchers and professors have been teaching and graduating some of the brightest students and most outstanding talents in the world.

One of the principal reasons behind this is the lifetime work of Geoffrey Hinton at the University of Toronto. Deemed by many to be "the godfather of deep learning" for his pioneering work over many years at this university, Hinton has worked for Google since 2013 in both California and Toronto. He returned to Toronto from California recently to head up the new Vector

Institute for Artificial Intelligence, an independent, non-profit organization affiliated with the University of Toronto that is receiving strong support from governments, large corporations, and private benefactors.

As a result of developments in this field over the last three decades, Canada has not only been producing some of the brightest minds and most creative AI talents, researchers, and teachers in the world—many of whom have been snapped up by the United States and other countries—but Canada has benefited immensely from Donald Trump's decision to close the U.S. border to highly skilled and talented immigrants from specific countries who have ended up coming to Canada and profiting from technological opportunities and applying their expertise here. As a result, Toronto, Montreal, Calgary, and Edmonton are beehives of activity in this field thanks to educators like Yoshua Bengio, one of the co-founders of deep learning, and the Montreal Institute for Learning Algorithms, as well as Richard Sutton, a professor of computer science and iCORE Chair at the University of Alberta, who is considered by some to be one of the real leaders in modern computational reinforcement learning and has made innovative contributions to reinforcement learning and temporal difference learning that are being acknowledged and used in other parts of the world.

It is developments like these that explain why LinkedIn's founder Reid Hoffman donated $2.45 million in 2018—the largest donation ever made to the University of Toronto's Faculty of Information—to create a new chair to investigate the relationship between artificial intelligence, human beings, and humanity in general. As Hoffman stated when this donation was made, "Artificial intelligence will revolutionize how we live, creating both incredible opportunities for benefits, as well as some disruptions that will be important to manage. The future is always sooner and stranger than you think, and so while it's difficult to predict exactly where this technology will take us, it is hugely important to focus some of our best minds on the implications of living with AI. I'm delighted the Chair will focus on these enormous questions at the University of Toronto's iSchool."

The donation by Reid Hoffman was followed by a donation of $100 million from the Gerald Schwartz and Heather Reisman Foundation to the University of Toronto in 2019 to further research activities related to artificial intelligence, biomedicine, how new technologies can both disrupt and enrich people's lives, as well as ethical issues surrounding and related to AI. Part of this donation will go towards constructing a new 750,000-square-foot complex at the northeast corner of College Street and Queen's Park that will be called the Schwartz Reisman Innovation Centre in phase one and the Institute for Technology and Society in phase two. It is the largest single donation that has been made to the University of Toronto throughout its history, thereby attesting to the great importance of this area and the key role Canadians and Canadian creativity are expected to play in its development in the future.

Creative contributions by Canadians to artificial intelligence and related areas throughout the world are not limited to educational institutions or the theoretical side of these fields. There have been some highly creative contributions on and to the practical side as well. For example, Canadian Sandra Mau, with a bachelor's degree in engineering science (aerospace) from the University of Toronto, a master's degree in robotics from Carnegie Mellon University in Pittsburgh, and a master's degree in business administration from Queensland University of Technology in Australia, is the founder and CEO of a company called TrademarkVision. This rapidly expanding and extremely successful company has created one of the world's most effective and frequently used visual search engines for trademark brands and logos based on proprietary image recognition, identification, and protection. It is already being used by many of the major governments, corporations, and law firms in the world to document, record, and compare trademark brands and logos in order to protect intellectual property rights. This company was acquired by Clarivate Analytics in 2018, thereby making it possible for TrademarkVision to expand its operations to all parts of the world.

Like Canada's educational system, the creation and development of the country's political system has also been long and arduous. It can be traced back to the Indigenous people, and to the creation of "confederacies" to provide effective government, governance, and leadership for the communities and societies they served. The best-known examples of this are the Huron Confederacy and the Iroquois Confederacy.

The Huron Confederacy was created in central Ontario in an area around Georgian Bay and Lake Huron. It brought together four major bands and 20 autonomous villages through the establishment of a governing council chosen by clan mothers that provided procedures and policies for settling disputes and collaborating on matters of mutual interest and concern. Even better known was the Iroquois Confederacy, which straddled eastern Ontario and upper New York State. It resulted from the creation of the Five Nations League and a Grand Council composed of 50 federal or national chiefs. Most matters were resolved peacefully through discussions, debates, compromises, concessions, and the passing of the peace pipe. This is seen by some as a model for political and governmental decision-making—a model that is said to have been used to advantage when the Constitution of the United States was drawn up.

Also important in a political sense was the decision by Louis XIV in France to revoke the charter given to the Company of the Hundred Associates and make New France a royal province in 1663. This had a fundamental bearing on virtually everything that has followed since that time. Giving New France the same status as a province in France did more than anything else to establish a permanent and powerful French-speaking colony in North America, thereby giving rise to expectations about how this colony was to be governed and treated politically and culturally that have prevailed to the present day.

With the passage of the Royal Proclamation in 1763 that prescribed boundaries, government, laws, and regulations for British holdings in North America, it was assumed that the French would be assimilated into the British system of politics, government, education, and overall way of life. It was not long, however, before it was apparent that this approach was not working. Not only was the culture and colony of New France too strongly entrenched to be swept aside or subjected to British rule, but also Britain required the support of the French colonists in dealing with the rapidly expanding, ambitious, and discontented English-speaking colonies—the famous Thirteen Colonies—along the Atlantic seaboard of what is now the United States.

As a result, the British Government passed the Quebec Act in 1774. It proved to be a far more creative response to an exceedingly difficult and complicated political problem. Not only did it recognize that homogeneity was not possible in the northern half of the North American continent, but also it signaled the beginning of "duality" between the French and the English in Canada. It did so by placing much more emphasis on integration and understanding of cultural differences than on assimilation, oppression, and coercion. This enabled the French to preserve many of their cherished legal and civic procedures, cultural practices, and overall way of life. Regrettably, the Quebec Act caused a great deal of resentment at that time despite its highly original and creative character. English colonists and settlers in what became Canada almost a hundred years later felt that the French were given too much power and authority over their affairs by the British Government and the large amount of land granted to Quebec was totally unacceptable. Elsewhere, the Quebec Act was included in the list of "Intolerable Acts" by the Thirteen Colonies to the south that led to the American Revolution and eventually the creation of the United States as an independent and sovereign country.

Despite this, the Quebec Act did an enormous amount to shape the political future of Canada. It did so by recognizing the dual character of British North America and the distinctiveness of the French-speaking population. This made respect for cultural differences—rather than subservience to a single dominant English culture—the cornerstone of Canada's rapidly evolving political system. While the British government did this for strategic reasons—it didn't want to be involved in a war with the French colony at the same time that it was dealing with problems with the colonies to the south—it had a profound impact on the overall development of Canada. Not only did it secure the existence of French Canada and Quebec as distinctive entities in Canada—a "distinct nation," as it is referred to following official recognition of this by the government of Canada—but also it gave rise to an equally distinct and loyal English Canada, as well as the creation of a comprehensive framework within which relations between the French and the English in Canada would be conducted and hammered out in the future. In so doing, it opened the doors for Canada to eventually become a bilingual, multicultural country, with two official languages and many diverse cultures.

If the Quebec Act had a powerful effect on the development of Canada and its political institutions and system, so did the War of 1812. This is especially true for the battles at Queenston Heights in 1812, and Chrysler's Farm and Chateauguay in 1813. Had the War of 1812 not been won, it is very likely that Canada would be part of the United States today. But there is another important reason why the War of 1812 was a major political and military accomplishment. It was a war of defence and survival rather than a war of rebellion and aggression. As a result, it did a great deal to shape the conviction among Canadians that they are prepared to fight whenever this is required but they do not initiate wars unless it is necessary to do so. This has had a profound impact on Canada's political evolution, including the creation of the BNA Act and its commitment to "peace, order, and good government," the way Canadians approached the First and Second World Wars, foreign policy procedures, practices, and initiatives such as peacekeeping and peacemaking, and the decision by Canada not to participate in the invasion of Iraq.

While Canada came into existence as a country at Confederation in 1867, most Canadians and historians believe it "came of age" during the First World War. This is because Canada played a major role in this war from 1914 to 1918, as well as a crucial role in the final stages of the war, especially at Vimy Ridge. On Easter Day, April 9, 1917, a Canadian Corps of four divisions was assigned its first independent mission. It was to capture Vimy Ridge after all other Allied forces had failed to take this strategically located stronghold held by the Germans. The Canadian Corps succeeded brilliantly in this campaign after a powerful and highly creative offensive. They sustained this momentum by spearheading the final Allied campaign from August 4 to November 11, 1918. As a result, this period has often been called "Canada's finest hour" and "Canada's one hundred days."

These events constitute one of those rare defining moments in history that acquire greater political, military, and international importance over time. The reason for this is not difficult to detect. Canada was finally recognized throughout the world as a country in its own right—a country with strong historical and emotional ties to France and Great Britain but nevertheless separate from them and capable of acting on its own. For although Canada was given power over its external affairs when the BNA Act was passed in 1867, everybody knew that the British government continued to called the shots when it came to Canadian foreign policy.

It was only after the First World War that Canada was recognized in the world as a sovereign nation and independent political entity. Not only did it sign the Versailles peace treaty in its own right, but also it became an independent member of the League of Nations despite strong opposition to this from the United States. In 1932, the Statute of Westminster formally recognized the equality of Canada (and the other dominions) in the British Commonwealth of Nations. Canada played a leading role among these self-governing dominions, which also included Australia, New Zealand, and South Africa, in advocating for the Statute and its passage into law. Although the English monarch remained Canada's monarch as long as Canada so desired, all other involvements by the British government in Canadian affairs, unless specifically requested by Canada, came to an end.

It was around this time that a number of highly creative female "political trailblazers" made their mark on the Canadian and international political scene. Most prominent among these trailblazers, all of whom lived and worked in Alberta for a time, were Emily Murphy, Canada's and the British Empire's first female judge and the first woman appointed to the bench in the British Empire; Louise McKinney, who when elected to the Legislative Assembly of Alberta in 1917 became the first female elected official in the British Empire; Irene Parlby, who emigrated to Alberta from England in 1896 and became the first female cabinet minister in Alberta in 1921; Nellie McClung, who was elected to the Alberta legislature in 1921 and was a well-known social and political activist as well as an outstanding orator and author; and Henrietta Muir Edwards, who was instrumental in founding the forerunner of the YMCA in 1875 and helped to establish the National Council of Women in 1893 as well as being a founding member of the Victorian Order of Nurses.

While each of these women was highly creative and achieved an enormous amount in their own right, their greatest achievement was collective rather than individual. This occurred in 1928 when they petitioned the Supreme Court of Canada to recognize women as "persons" under Section 24 of the British North America Act of 1867. When the Court ruled against them, they took their petition to the judicial committee of the Privy Council in London, which was the court of last appeal at that time. The judicial committee overturned the decision of Canada's Supreme Court and ruled that women are indeed "persons" in the legal and legislative sense. In recognition of this achievement, and many others, these "famous five women" were appointed honorary Canadian senators some 80 years later in 2009. One wonders how many women throughout the world who have fought for women's rights and recognition as "persons" in the legal and legislative sense—or who have wanted to have careers in politics—have been inspired by the creative accomplishments of these five courageous Canadian women over the years! Their contributions were so substantial and heroic that statues commemorating their achievements are ensconced outside the East Block of the Parliament Buildings in Ottawa.

These women were not the only Canadians to make seminal contributions to key political developments that have had an important impact on the world. Three others are John Humphrey, Louise Arbour, and Philippe Kirsch.

John Humphrey was born in Hampton, New Brunswick in 1905. Unfortunately, he lost both his parents when he was young to cancer and also lost one of his arms in a fire. He was often teased by other boys at the boarding school he attended and many felt these experiences were major factors in building his confidence, character, and compassion. After teaching law and international relations at McGill University in Montreal and becoming deeply immersed in human rights, Humphrey got an opportunity to join the United Nations where he was eventually appointed first director of the United Nations Division of Human Rights. In this position, he wrote the initial draft of the Universal Declaration of Human Rights—a remarkable and highly creative achievement in its own right—and then worked closely with Eleanor Roosevelt on the finalization, affirmation, and implementation of this Declaration.

Humphrey remained at the United Nations for 20 years,

during which time he oversaw the implementation of 67 international conventions and the constitutions of dozens of countries. He also worked in such areas as freedom of the press, status of women, and racial discrimination at the UN, as well as travelling to many countries throughout the world to investigate and oversee developments in these areas. He was also a director of the International League for Human Rights, a member of the Royal Commission on the Status of Women in Canada, and was instrumental with others in creating the Canadian Human Rights Foundation. In 1988, he received the UN Human Rights Award on the fortieth anniversary of the passage of the Universal Declaration of Human Rights, a fitting tribute for his remarkable international contributions.

Like John Humphrey, Louise Arbour and Philippe Kirsch have made many creative contributions to legal affairs, justice, and human rights throughout the world, largely through the International Criminal Court and International Criminal Tribunal, which are responsible for prosecuting individuals for crimes of genocide, crimes against humanity, and war crimes.

Louise Arbour held many important academic, political, and diplomatic positions in Canada, including at the Osgoode Hall Law School at York University, the Supreme Court and Court of Appeal in Ontario, and the Supreme Court of Canada, before being appointed chief prosecutor of the International Criminal Tribunal for Rwanda in Arusha and the International Criminal Tribunal for the former Yugoslavia in The Hague. In this latter capacity, she indicted Yugoslavia President Slobodan Milošović for war crimes, the first time a serving head of state was called to account before an international tribunal or court, something which led to several other political authorities being indicted as well. In 2017, she was appointed Special Representative for International Migration by UN Secretary-General António Gutierres and, as such, is responsible for following up on the 2016 high-level Summit on Addressing Large Movements of Refugees and Migrants in the hope of concluding negotiations on a global compact on numerous migration and refugee matters. During her distinguished career, she has also authored many books and articles dealing with criminal procedures, criminal law, and international relations, including *War Crimes and the Culture of Peace, A Global Agenda*, and *The Global Refugee Crisis*.

What makes Arbour's contributions even more meaningful is the fact that another Canadian, Philippe Kirsch, who came to Canada from Belgium at a very young age, was a major pioneer in the founding of the International Criminal Court and Tribunal. This occurred in 1998 when Kirsch served as chairman of the Committee of the Whole of the Diplomatic Conference of Plenipotentiaries on the Establishment of an International Criminal Court. He was also chairman of the Preparatory Commission for the International Criminal Court from 1999 to 2002 and served as its first president from 2003 to 2009. In this capacity, he made many contributions to ensuring that international crimes of this type should never go unpunished and every effort should be made to bring perpetrators to justice.

For many years prior to the international legal contributions of Arbour, Kirsch, and others, many Canadians thought that Canada's greatest contribution to the world was peacekeeping, which is not surprising in view of the fact that the Canadian Constitution is predicated on "peace, order, and good government" as indicated earlier. The impetus for Canada's commitment to peacekeeping came not only from the Constitution, but also from the international achievements of

Lester Pearson, who played a leading and highly creative role in the UN intervention in the Suez Crisis in 1956. Not only was Pearson (who later became prime minister) instrumental in creating the UN Emergency Force that stabilized the Suez situation and facilitated the withdrawal of French, British, and Israeli forces, but also he was instrumental in organizing Canada's contribution of signals, reconnaissance, and supply units and to seeing that the entire mission was under the command of a Canadian, Major-General E.L.M. Burns. When Pearson was awarded a Nobel Peace Prize for his creative achievements in diffusing this volatile situation, it was easy for many Canadians to conclude that Canada had a major and legitimate role to play as a significant peacekeeper in the world.

As a result of developments like this, and others, Canada stepped up its involvement in peacekeeping initiatives that were designed to monitor the movements of warring factions and armies, supervise and negotiate ceasefires, facilitate the signing of peace agreements and treaties, and protect citizens. This included the international Commission for Supervision and Control of Vietnam, Laos, and Cambodia, and the dispatching of peacekeepers to the Congo in 1960, West (Papua) New Guinea in 1962, Yemen in 1963, and Cyprus in 1964. This led to more peacekeeping operations and Canada eventually sharing the Nobel Peace Prize to UN peacekeeping forces in 1988. It is estimated that between 1948 and 1988, Canada contributed more than 80,000 personnel to peacekeeping missions and operations in different parts of the world, a remarkable 10 percent of the total of all UN peacekeeping forces. Canada also contributed a significant portion of the UN forces that were sent to the Balkans under the command of Canadian General Lewis MacKenzie, especially in Sarajevo and Bosnia-Herzegovina.

Unfortunately, it has proven difficult to sustain this momentum over the last three decades. Things took a turn for the worse in 1992–93 when two Canadian soldiers beat a local Somali teenager to death during peacekeeping operations in Somalia. This resulted in a great deal of negative publicity for Canada and ultimately a high-profile public enquiry to investigate this deplorable event. Matters worsened in 1994 when Roméo Dallaire was force commander of UNAMIR, the ill-fated United Nations peacekeeping mission in Rwanda. As a result of these events, as well as several other developments that took place in other parts of the world around this time, peacekeeping fell out of favour in Canada and the government of Canada began to shift its attention elsewhere.

In 1995, Canada began to lessen its involvement in UN peacekeeping operations and increase its involvement in NATO activities, especially in countries like Afghanistan, Libya, and the Ukraine. What made this work especially difficult when it was commenced—and even more difficult today—is the fact that peace*making* is far more difficult than peace*keeping*. Under such circumstances, Canada finds itself in a very awkward position today, as evidenced by the recent difficulties in Syria, the many battles against terrorists in general and ISIS forces in particular, and other skirmishes in the Middle East and elsewhere in the world.

Over the past decade, many Canadians have felt that Canada's identity in the national and international sense has much more to do with multiculturalism and the country's official political policy in this area than it does with peacekeeping or peacemaking. While the roots of multiculturalism and the present policy of multiculturalism are buried deep in Canada's historical traditions and demographic situation as noted earlier—as well as in the need to welcome immigrants and refugees from other parts of the world to live and work in Canada—the multicultural policy as such came into existence in 1971 when the government of Canada declared that henceforth Canada would officially be "a bilingual, multicultural country." Paving the way for this declaration was the Royal Commission on Bilingualism and Biculturalism, which operated from 1963 to 1965. When the Royal Commission was established, the prevailing assumption was that Canada was primarily a bilingual, bicultural country with two main "founding peoples and cultures" and two basic languages, French and English.

During the many discussions and debates that took place in connection with the commission, it became clear that the Indigenous peoples of Canada and most Canadians who were neither of English nor French descent were prepared to accept the fact that Canada was a *bilingual* country but they were not prepared to accept that Canada was a *bicultural* country. They claimed that the Indigenous peoples and many ethnic groups other than English and French had made—and were making—indispensable contributions to the development of Canada that should be officially recognized and given their due. Recognition of this fact eventually caused the commission to replace its description of Canada as a bilingual and bicultural country with its characterization of Canada as a bilingual, *multicultural* country.

When the commission released its final report in a series of publications between 1965 and 1970, Book IV, which was released in 1969, was devoted almost entirely to the contributions made by people, groups, and communities other than English and French. The Official Languages Act was also passed in that year. It declared that French and English would be the two official languages of Canada. As such, all federal institutions would be compelled to provide their services in either English or French depending on citizens' choice. The Act also created a Commissioner of Official Languages to oversee the Act and enforce its regulations and requirements.

This Act, together with the work undertaken by the Bilingualism and Biculturalism Commission, paved the way in 1971 for the government of Canada to declare that Canada would from this time forward be a bilingual, multicultural country with "multiculturalism within a bilingual framework" as its official policy. The government went on to state that "we believe that cultural pluralism is the very essence of Canadian identity. Every ethnic group has the right to preserve and develop its own culture and values within the Canadian context. To say that we have two official languages is not to say we have two official cultures, and no particular culture is more official than another. A policy of multiculturalism must be a policy for all Canadians."

The government further stipulated that public support and encouragement would be provided to all the "various cultures and ethnic groups that give structure and vitality to our society. *They will be encouraged to share their cultural expression and values with other Canadians and so contribute to a richer life for us all.*" In stating this, the federal government was careful to emphasize that there are two fundamental and essential dimensions to multiculturalism in Canada: providing opportunities for Canada's diverse cultural and ethnic groups to preserve and develop their own cultures and cultural customs, traditions, and identities; and sharing their cultures and cultural customs, traditions, and identities with others. This helps to ensure that cultures and ethnic groups do not become isolated from one another or insular, but rather create fundamental connections, interactions, interrelationships, and exchanges between their cultures and all the other cultures and ethnic groups in the country. The fact that both aspects of multiculturalism are imperative for it to work successfully goes a long way towards explaining why multiculturalism and cultural pluralism have worked well in Canada on the whole and better than they have worked in many other countries in the world.

The federal government has maintained and reinforced this policy of multiculturalism and bilingualism ever since it was officially declared in 1971. It has done this through a variety of instruments, vehicles, and funding programs, such as the appointment of a Minister of State for Multiculturalism and, later, the formation of a Ministry of Multiculturalism and Canadian Identity; rigorous enforcement of bilingualism in all federal institutions; confirmation of the political policy of bilingualism and multiculturalism in the Canadian Constitution and Charter of Rights and Freedoms of 1982; passage of the Canadian Multiculturalism Act in 1988; specially designed funding programs to enable cultural and ethnic groups to share cultural experiences with others; and many additional initiatives and activities. While there has been some resistance to these pronouncements, enactments, and programs in certain quarters, by and large the policy of multiculturalism in a bilingual framework has been endorsed and embraced by all governments in Canada—municipal, provincial, and territorial in addition to federal—as well as countless organizations and institutions in the private sector and the large majority of Canadians.

While many more people speak languages other than English

and French in Canada today than was the case in 1971, there is no doubt that Canada's bilingual and multicultural policy accurately describes the reality of the Canadian situation at present, with English and French as the two main languages and many different cultures in existence throughout the country. Indeed, Canadian society is becoming increasingly multicultural all the time with the arrival of many more immigrants and refugees from other parts of the world. The OMNI television system is a perfect illustration of this. With many stations across the country, it broadcasts in more than 30 languages and is now the largest producer of multicultural programming in the world. There is also a much larger commitment to multiculturalism, interculturalism, cultural diversity, and pluralism in communities and educational institutions across the country than there was in the past.

This explains why Canada was one of the initiators and principal formulators—as well as one of the strongest advocates—of UNESCO's Universal Declaration for the Protection and Promotion of the Diversity of Cultural Expressions, which was signed in 2005. The impetus for this began in the late 1990s when Canada's heritage minister, Sheila Copps, convened an international conference of minsters of culture from around the world. This conference was designed to discuss the threat to cultural diversity and pluralism that was evident as a result of the international spread of mass media under the control of a few very powerful producers and countries, especially the United States.

The two countries that were most concerned with this problem were Canada and France. As a result, steps were taken in Canada in collaboration with authorities in France to prepare, promote, and develop an international legislative document that would protect cultural diversity in all countries of the world, thereby enabling people to enjoy their own cultural creations as well as those of other countries. Not only were Canadians actively engaged in the original creation of this document, but Canada was the first signatory of the UNESCO declaration and has been at the cutting edge of international efforts to promote and implement the declaration in all parts of the world since that time. Not surprisingly, the International Federation of Coalitions for Cultural Diversity has its headquarters in Canada for these very reasons.

As a result of achievements such as these, there is growing awareness in Canada that the country's principal contribution to the world at present is its ability to welcome people from other parts of the world as well as its official policy of multiculturalism. As revealed in the 2018 Canada's World Survey conducted by the Environics Institute for Survey Research, the Canadian International Council, SFU Public Square, and the Bill Graham Centre for Contemporary International History, Canadians generally see Canada as a positive force in the world in these areas:

> Multiculturalism, diversity and inclusion are increasingly seen by Canadians as their country's most notable contribution to the world. It is now less about peacekeeping and foreign aid, and more about who we are now becoming as a people and how we get along with each other. Multiculturalism and the acceptance of immigrants and refugees now stand out as the best way Canadians feel their country can be a role model for others, and as a way to exert influence on the global stage. Moreover, Canadians are paying greater attention

to issues related to immigration and refugees than they did a decade ago; their top interest in traveling abroad remains learning about another culture and language; and they increasingly believe that having Canadians living abroad is a good thing, because it helps spread Canadian culture and values (which include diversity) beyond our shores.

The survey goes on to say that Canadians feel that a neighbourly immigration policy is the best way for the country to exert its influence and make its contribution internationally. According to Michael Adams, president of the Environics Institute, "Canadians cite our traditions and policies of bilingualism, multiculturalism and mutual accommodation as key achievements. They also note our acceptance of immigrants and refugees from around the planet." Canada's experiences with immigrants and refugees from Vietnam, South America, and more recently Syria and other parts of the world bear this out in many ways and have been a major factor in this regard.

There is one other major Canadian political milestone that must be addressed here because it is highly creative in nature and has profound international implications. It is the signing of the Constitution Act in 1982. Not only did this bring the Canadian constitution home from Britain, but also it guaranteed certain fundamental rights and freedoms for all Canadian citizens.

No one was more responsible for this milestone than Pierre Elliott Trudeau. As prime minister, he was without doubt the driving force behind the preparation and signing of the Constitution Act of 1982, which enabled patriation of the BNA Act of 1867 and its subsequent amendment, as well as the passage of the Charter of Rights and Freedoms. As Trudeau said on April 17, 1982, the day the Constitution was signed, "Today, at long last, Canada is acquiring full and complete national sovereignty. The Constitution of Canada has come home. The most fundamental law of the land will now be capable of being amended in Canada, without any further recourse to the Parliament of the United Kingdom." He went on to say:

> I speak of a Canada where men and women of Aboriginal [Indigenous] ancestry, of French and British heritage, of the diverse cultures of the world, demonstrate the will to share this land in peace, in justice, and with mutual respect. I speak of a Canada which is proud of, and strengthened by, its essential bilingual destiny, a Canada whose people believe in sharing and in mutual support, and not in building regional barriers. I speak of a country where every person is free to fulfill himself or herself to the utmost, unhindered by the arbitrary actions of governments.

In total, the Constitution Act is comprised of seven main sections. In addition to incorporating the BNA Act (now renamed the Constitution Act) and the Charter of Rights and Freedoms, it includes specific sections dealing with the rights of the Aboriginal (Indigenous) peoples of Canada, equalization and regional disparities, the powers and responsibilities of the federal, provincial, and territorial governments, constitutional conferences, a procedure for amending the constitution, an amendment to the 1867 Act dealing with "non-renewable natural resources, forestry resources and electrical energy," and a general section dealing with the overall intentions, interpretation, and implementation of the Act. However, the bulk of the Act has to do with ensuring that every Canadian citizen is accorded and guaranteed certain basic human rights and freedoms, including freedom of conscience, religion, thought, belief, opinion, expression, the press, peaceful assembly, and association. The Act and Charter also spell out Canada's commitment to democratic, mobility, legal, and equality rights, as well as the official languages of Canada and minority language rights in education. There is also a section in the Charter that includes a "notwithstanding clause," which enables legislatures to override rights in extreme cases and situations. For the most part, great political restraint has been exercised with respect to the use of this safety valve.

Patriation of the Constitution and passage of the Charter of Rights and Freedoms were exceptionally difficult to achieve and required a great deal of debate, disagreement, concessions, and compromises. Even after all this, the outcome left something to be desired since René Lévesque, premier of Quebec at that time, refused to sign the accord because he felt he had been betrayed by the other provincial premiers and Quebec had been "abandoned at the very moment of crisis." As a result, there is still an element of incompleteness to Canada's constitutional arrangements that remains to be resolved.

Nevertheless, it is still a comprehensive and ingenious document that goes well beyond the constitutions of virtually all other countries in terms of enshrining the basic rights and freedoms of citizens, thereby making it of great historical and contemporary relevance and value to all people and countries and the world as a whole. In keeping with the ideals enshrined in the Charter of Rights and Freedoms is the recently created Human Rights Museum. It is located in Winnipeg and recognizes the achievements of Canadians in this field. Canada is one of the first countries in the world—if not *the* first—to create a major museum of this type.

As Richard Gwyn, one of Canada's most distinguished journalists, said shortly after the signing of the Constitution Act and the passage of the Charter of Rights and Freedoms:

> My strongest first impression is that Canadians, without really realizing it, are engaged in a social experiment for which there is no real equal anywhere in the world. It's one that only an unusually tolerant and civil people could undertake.
>
> It's a "rights revolution." It is exhilarating. It is also scary. It has the potential to be extraordinarily creative. It could end up in chaos, even in self-destruction.
>
> What is happening here, as I perceive it, is a social experiment in the articulation of and the realization to the maximum extent possible of individual rights and, even more so now, of group rights.
>
> We are refining the meaning of citizenship in order to accommodate—to provide needed space for would be a better phrase—all kinds of equality groups, women, gays, blacks, native people, ethnic groups, the handicapped and, in a different sense, also for all kinds of advocacy groups from environmentalists to non-smokers.[15]

Many people throughout the world have expressed their admiration of Canada for having the courage, determination, and commitment to create and pass the Charter of Rights and Freedoms. For what is at the heart of this Charter is the conflict that can and often does exist in both theory and practice

---

15. Richard Gwyn, "Canada's 'Rights Revolution,'" *Toronto Star*, July 19, 1992, pp. B1, B8.

between the collective and political rights of the state and the individual and personal rights of citizens and citizen groups. When the Charter was signed, many feared that it would lead to endless debates and countless court battles and cases between the country's governments on the one hand and citizens and citizen groups on the other hand. However, in retrospect, this has not happened, at least to the degree that many expected.

The implications of this legislation are considerable for many countries in the world because they indicate that charters of this type may be one of the best means for reconciling the differences, debates, conflicts, and even violence that can result or erupt in countries between the powers and rights of governments and countries and the powers and rights of citizens and local, regional, and national groups.

According to recent research undertaken by law professors David Law of Washington University in Saint Louis and Mila Versteeg of the University of Virginia, "a stark contrast can be drawn between the declining attraction of the U.S. Constitution as a model for other countries and the increasing attraction of the model provided by America's neighbour to the north, Canada." They go on to state that Canada is rapidly becoming a "constitutional superpower among common law countries and is serving as a trendsetter." This is true not only of the Canadian Constitution—as Nelson Mandela said in a speech to the Parliament of Canada, "Your respect for diversity within your own society and your tolerant and civilized manner of dealing with the challenges of difference and diversity had always been our inspiration"—but also of the Charter of Rights and Freedoms, which has been described as the leading influence on Israel's basic laws as well as bills of rights passed in Hong Kong, South Africa, and New Zealand.

What makes Canada an ideal case study and valuable model in this regard is the fact that many of the activities of activists and advocates in Canada since the passage of the Charter of Rights and Freedoms have kept Canada in the forefront of political legislation and social change throughout the world. In fact, many of the social transformations that have taken place in Canada since 1982 can be traced back to the passage of the Charter of Rights and Freedoms in one form or another, such as limiting the powers of the police, clarifying and strengthening the reproductive rights of women, recognition of LGBTQ communities, protection of linguistic rights and the rights of francophones outside Quebec, the expansion of Indigenous (Aboriginal) rights, and what is called "judicial activism" or the transfer of policy-making to the courts, especially with respect to morality issues.

A few examples should suffice to illustrate Canada's experiences and achievements in this area—experiences and achievements that often start with illegal activities or initiatives by one or two courageous citizens or small groups, gather momentum when other citizens join the causes that they represent, and eventually end by becoming law. Most prominent in this regard are political endeavours or social developments with respect to abortion, same-sex marriages, assisted death, and the legalization of marijuana. All of these examples, and others, have involved citizens who have defied the law in the beginning but made such a strong case for their particular cause or issue that many other citizens joined the ranks to demand passage of legislation affecting all Canadians.

Consider abortion as one example of this. At the present time, abortion is legal in Canada at all stages of pregnancy and is governed by the Canada Health Act. While some non-legal obstacles still exist, Canada is one of a handful of countries in the world that has no *legal* restrictions on abortion. This has not always been the case. In fact, abortion was illegal in Canada as recently as 1969. Things started to change and change substantially, and even violently, in this regard when Dr. Henry Morgentaler began to provide abortions in his clinic in defiance of the law. Very quickly, Canadians were divided into two major groups on this issue: "pro-life" advocates who believed that abortion was wrong and should be prohibited under most or all circumstances, and "pro-choice" advocates who felt that abortion was a personal matter that should be decided by pregnant women themselves and not by the state. While this issue has continued to simmer and divide Canadians to the present day on an almost 50–50 basis, the 1969 law was struck down by the Supreme Court of Canada in 1988 on the basis that it violated a woman's right to "life, liberty, and security of the person," as guaranteed in Section 7 of the Charter of Rights and Freedoms. Abortion was and still is a very contentious issue in Canada. However, most Canadians seem prepared to leave the issue where it stands at present now that abortions are legal. Moreover, most governments are equally reluctant to confront this issue because it would be politically divisive or explosive to do so.

Like abortion, same-sex marriages have also been a very contentious and divisive issue in Canada, probably because the idea of same-sex marriage, like abortion, carries deep religious connotations, strong social overtones, and serious political consequences. Here, as well, Canada played a creative role in coming to grips with this issue. It was the fourth country in the world—and the first outside Europe—to legalize same-sex marriage on a nationwide basis following its legalization in the

Netherlands, Belgium, and Spain. This occurred on July 20, 2005 with the enactment of the Civil Marriage Act, following years of calls to make lesbian and gay marriages legal. Soon after this, there was a flood of same-sex marriages in Canada; this seemed to confirm that many Canadians felt this was the right thing to do even if it did not affect them.

Assisted death is another area with deep religious connotations as well as strong social overtones and political consequences. Here again, Canada is one of the international leaders in this field, after the Netherlands, Colombia, Luxembourg, and Belgium. Voluntary active euthanasia, called "physician-assisted dying," is now legal in Canada for all people over the age of 18 who have a terminal illness that has progressed to the point where natural death is "reasonably foreseeable." Legalization of the practice came as a result of a series of Supreme Court rulings striking down Canada's ban on medically assisted suicide. On June 17, 2016 a bill to legally allow assisted death within Canada became law. This time it was a province—Quebec—rather than a citizen, group of citizens, or activist that took the initiative to have assisted death legalized in Canada. Over 4,000 Canadians have chosen medically assisted death to end their lives since this practice became legal in 2016. While statistics are still very sketchy on this very recent matter, it would appear that generally speaking roughly 40 percent of these deaths occur in hospitals, 40 percent in homes, and the rest elsewhere; most people who choose medically assisted death are between the ages of 55 and 90 with the average age being in the low seventies; approximately half are women and half are men; and over 60 percent of cases result from cancer, followed by circulatory and respiratory illnesses and neuro-degenerative disorders in that order.

Finally, and most recently of all, Canada became the second country in the world to fully legalize the consumption and use of marijuana. This occurred on October 17, 2018 when the federal government made the consumption and use of cannabis legal for both medicinal and recreational purposes through the passage of the Cannabis Act. In this case, a few dedicated citizens, activists, and advocates challenged the law by selling or growing marijuana, followed by an increasing and substantial number of Canadians who became engaged in this practice or committed to this cause. After many consultations and assessments, the government decided to legalize the consumption, possession, and sale of restricted amounts of marijuana or cannabis and left it largely to the provincial governments to decide whether it would be produced, dispensed, and sold by public and/or private companies devoted to this purpose. Concerted attempts are also being made at the present time to pardon the people who were involved in these activities when they were illegal. The problem now is how to regulate and control the production, use, and sale of this substance so that black-market activities are reduced or hopefully eliminated, under-age children are protected, and consumption of this drug does not get out of hand.

Canada's political impact on the world is not confined to specific matters such as these. It also includes Canada's impact on the world in more general terms. According to a recent survey by Ipsos MORI, the United States is increasingly being disregarded as having a positive influence on the world. This survey polled 18,000 respondents in 24 countries and it found that only 40 percent of people think the U.S. has a strong or somewhat positive influence on the world. That's less than China, whose global influence is viewed favourably by 49 percent of people, although this is well ahead of Russia at 35 percent. Canada, on the other hand, was seen as setting the best example of all, with 81 percent of respondents saying Canada has a positive influence and effect on global issues and world affairs. The poll's numbers confirm the fact that many of Canada's most significant and creative achievements have had a beneficial effect on the world and are changing the world in some fundamental and very positive ways.

Chapter Nine
Looking Back, Going Forward

More than a quarter of a century has elapsed since I sat down on that cold evening in late November to begin reading the *Canadian Encyclopedia*. During the long and tedious winter that followed, I worked my way through the entire encyclopedia, stopping to make notes on countless entries that were of great interest to me and I wanted to record for future use, but also scanning some other entries that were of less interest.

Working my way through the *Canadian Encyclopedia* from beginning to end was one of the best things I have ever done in life. Not only did I learn an enormous amount about Canada, Canadians, and Canadian culture, but also it provided a comprehensive perspective and detailed understanding of Canada's historical and contemporary development in many different ways.

I began to realize how many sacrifices past generations of Canadians made to improve life for themselves, their families, and future generations. This included exploring and settling a huge expanse of land from the Atlantic Ocean in the east to the Pacific Ocean in the west and the Arctic Ocean in the north, searching every nook and cranny of that land looking for natural resources and basic foodstuffs, clearing millions of acres of land of rocks, boulders, trees, tree stumps, and brush, dredging countless swamps, working 12 to 14 hours a day with little in return, creating myriads of agricultural and industrial products, overcoming the manifest destiny movement and turning back several invasions from the United States, and fighting two world wars and losing more than a hundred thousand friends, family members, and neighbours in the process.

And this is not all. I also learned a great deal about the remarkable contributions that past generations of Canadians made to creating a sovereign and independent country despite incredible odds, developing one of the finest medical and health care systems in the world, winning numerous Nobel prizes, establishing many valuable economic, political, social, and educational institutions, and producing an array of world-famous artists, scientists, scholars, athletes, and many other types of people.

Had it not been for the sacrifices, contributions, and achievements of past generations of Canadians, present generations would not be able to enjoy the high standard of living and precious quality of life that exists throughout the country today. Nor would they be able to live longer, healthier, and more enjoyable lives, travel to many different places in the country and the world, and look forward to their retirement years with hope, optimism, and enthusiasm rather than pessimism, anxiety, and apprehension. For despite all the difficult problems that exist in Canada and the world today, most Canadians are far better off at present than they ever were in the past.

Matters like this have occupied my mind for more than 50 years now. As my work in this area has intensified and expanded, I have come to realize what factors have been most important in enabling Canadians to come to grips with the many challenges they have been confronted with as well as the cornucopia of unique opportunities they have been presented with over the centuries.

The challenges have resulted from the colossal size of the country, many exceedingly difficult transportation and communications problems, the cold climate, the dominance of winter, the diversity of people and languages, and the need to create a vital and viable country in the northern half of the North American continent. The opportunities have emanated from the munificent supply of natural resources, the superb location of Canada facing Europe in the east, Asia in the west, and the United States, the Caribbean, and Latin America in the south, millions of fresh water lakes and gushing streams and rivers, thousands of kilometres of beautiful coastline, excellent harbours, bays, and inlets, serene landscapes and seascapes, dense forests, and vast wilderness areas.

In order to come to grips with these challenges and take advantage of these opportunities, Canadians have had to be very creative. This has been necessary to survive in and develop a country that stretches across 5½ of the world's 24 time zones, is the second largest country in the world in geographical area, and has one of the highest standards of living and quality of life in the world. If "necessity is the mother of invention," surely it has manifested itself profusely in the Canadian case.

What stands out most clearly about the country's creativity is its expansiveness, pervasiveness, and variety. It manifests itself in virtually every aspect and dimension of the country's cultural life, from food, clothing, shelter, transportation, communications, agriculture, and industry to technology, the arts, sciences, sports, recreation, education, politics, and the natural environment. It also stems from all sectors and segments of Canadian society, from women and men, tiny towns and rural areas as well as sprawling cities, all the various provinces, territories, regions, and ethnic groups that exist throughout the country, and recent immigrants and refugees as well as long-time residents and well-established families.

Much of the country's creativity either occurred in Canada first, or in other countries in the world but as a result of inventions, artifacts, technologies, and products created by Canadians. While the impact of this creativity has been limited to Canada and Canadians in some cases, in most cases it has affected people and countries in many other parts of the world. This is one of the most fascinating things about Canadian creativity. In the process of coming to grips with their problems, challenges, opportunities, and possibilities, Canadians have created many works, processes, devices, and even entire fields that have not been confined in their geographical scope or impact to Canada, but have had a powerful effect on the whole world.

It is this fact that makes it possible to contend that Canadian creativity has not just influenced, improved, or contributed to the world, but has actually changed the world and changed it for the better in a whole series of profound and very effective ways.

There are many example of this, such as the invention of the telephone, paper from pulp wood, the electric light bulb, the electric oven, the screw propeller, Marquis wheat, basketball, lacrosse, standard time, the discovery of insulin, and highly creative contributions to automation, the emergence of modern medicine, and the mapping of vast geographical areas and the human brain.

To this list should be added seminal achievements in adult, distance, and early childhood education, key advances in the development of the radio, the creation of the snowmobile, the BlackBerry, the concept of soundscapes and the discipline of acoustic ecology, the birth and evolution of Hollywood, helping millions of people with mental and physical disabilities, the drafting of the Universal Declaration of Human Rights and the founding of the International Criminal Court and Tribunal, the realization of people's rights and freedoms, and a great deal else.

While creativity has never been a problem in Canada, what has been a problem, and a major problem indeed, is "innovation." While this is seldom discussed, there is a fundamental difference between the two. Whereas creativity is concerned with making things for the first time or making key changes in the character of things after they are produced, innovation is concerned with the practical application of creativity and the realization of financial, economic, commercial, and social benefits and opportunities that arise from creativity. This is where Canada and Canadians have consistently come up short, either because they have not had the financial resources, risk capital, advertising acumen, or markets, or are deemed to be too conservative or timid in nature to take advantage of the country's rich vein of creativity. As J. J. Brown pointed out in his insightful book *Ideas in Exile* many years ago:

> Time after time we have gotten there first after a magnificent sprint, and then stood around idly for years waiting to collect the risk capital required to get an industry going. Usually by the time we have solved the financial problem, other less torpid nations have caught up with and passed us. This happened with the variable-pitch propeller, with the hydrofoil boat, with the Jetliner, with the automatically controlled machine tools, with the electronic organ. Our inventors presented us with the world's first and a clear head start. But because of our conservative bent ... we were unable to do anything with it.[16]

16. J. J. Brown, *Ideas in Exile: A History of Canadian Invention* (Toronto: McClelland and Stewart Limited, 1967), p. 340.

Given this fact, the challenge of the future is clear and unequivocal. It is to focus much more attention, effort, and resources on generating the financial resources, risk capital, entrepreneurial expertise, start-up capabilities, and especially the markets that are necessary to capitalize on Canada's remarkable treasure trove of creativity. This means directing much more domestic and international investment into the practical application of creativity, the creation of many more research chairs, entrepreneurial programs, and innovative opportunities in the country's educational institutions, developing markets in all parts of the world and not just in some specific parts of the world, and stimulating creative activities designed to produce lucrative, cumulative, and long term results. It also means creating many more discovery districts, innovative centres, collective clusters, and dynamic hubs in towns, cities, and communities all across the country.

A good example of this is MaRS in Toronto. It is bringing together creative people in science, health care, communications, technology, and marketing to work closely with entrepreneurs, planners, policy-makers, corporate officials, civil servants, and politicians to exploit the opportunities that exist at present and are likely to exist in far greater numbers in the future. This includes a major university, five of the city's most prominent research hospitals, many global and business leaders, men, women, and members of different ethnic and cultural backgrounds, as well as creative start-ups capable of producing breakthroughs in medical research, the sciences, digital and artificial technology, social innovation, artistic creation, and industrial output.

What is true for MaRS is also true for discovery districts, collective clusters, and dynamic hubs in other Canadian cities, such as Montreal, Edmonton, Calgary, Vancouver, and Halifax. There is no reason why developments like this could not be generated in many other locations and places in Canada, thereby creating the collaborative possibilities and cooperative arrangements that are essential to overcome one of the biggest problems of all in the development and application of Canadian creativity in the years and decades ahead.

Of great importance in this regard is the need to invest much more heavily in the arts and especially arts education. It is a well-known fact that the arts, arts education, and creativity go hand in hand and are intimately connected. This is confirmed by the fact that more and more Canadian artists and arts organizations are achieving international reputations for their creativity and outstanding performances, exhibitions, and presentations throughout the world. The more this is facilitated, nurtured, promoted, and enhanced, the more younger generations of Canadians will be able to maintain, expand, and enrich the incredible legacy of creativity that past and present generations of Canadian artists and arts organizations have demonstrated and bestowed on the country and the world as a whole.

This is not only a case of unlocking the rich creative potential that exists in the arts and arts education across the country, but also unlocking the creative potential that exists in all Canadians and all sectors of Canadian society. It doesn't take a psychic to predict that the world of the future will be very different than the world of the present and past. Given all the technological changes that are going on in the world today, it is clear that economic opportunities, employment possibilities, and income prospects will be very different in the future than they are at present or have been in the past. Without the ability to create jobs and employment opportunities that are full-time rather than part-time, precarious, and contractual, many Canadian will find themselves out of work or confronted with dismal financial prospects. The only solution to this is to develop and use the creative potential that Canadians possess in abundance to a much greater extent.

In a democratic country like Canada, it is understandable that Canadians will manifest and develop their creativity in different ways as well as in many diverse areas and fields of the country's cultural life. This explains why creativity has ranged far and wide in Canada over the centuries. However, given the present state of the world and prospects for the future, it is hoped that a sizeable portion of Canadian creativity will be directed into coming to grips with all the difficult, demanding, and dangerous problems that exist in the world today. Most notable here are climate change, global warming, and the environmental crisis, huge disparities in income and wealth between rich and poor people and rich and poor countries, conflicts between different races, religions, ethnic groups, countries, cultures, and civilizations, the escalation of prejudice, inequality, injustice, hate, and violence in the world, and the perpetual threat of nuclear, chemical, and biological warfare.

These are all areas where Canadian creativity has manifested itself significantly in the past and is capable of manifesting itself even more so in the future. The Canadian capacity for caring, sharing, compassion, compromise, and cooperation is of utmost importance here, especially in terms of achieving environmental sustainability, reducing disparities in income and wealth, and improving the health and well-being of people and countries in all parts of the world and the world as a whole. This requires creativity in such areas as peacekeeping, peacemaking, multiculturalism, realizing unity in diversity, creating cultural exchanges and relations between the diverse peoples, races, religions, countries, and civilizations of the world, and especially making it possible for all people and all countries to enjoy reasonable standards of living and a decent quality of life without straining the world's scarce resources and finite carrying capacity to the breaking point.

The time has come to make a powerful commitment to ideals such as this—ideals that few countries in the world are as well positioned or endowed to achieve as Canada. Is this too much to ask for or expect? Surely Canada and Canadians are capable of making as many creative contributions to the world of the future as they have in the past and are at present. To do so would be to make an indispensable contribution to global development and human affairs at a crucial time in world history.

Selected Readings

Brown, J. J. *The Inventors: Great Ideas in Canadian Enterprise* (Toronto: McClelland and Stewart, 1967).

Brown, J. J. *Ideas in Exile: A History of Canadian Invention* (Toronto: McClelland and Stewart, 1967).

Canada Heirloom Series. *Wayfarers: Canadian Achievers, Vol. V* (Mississauga, ON: Heirloom Publishing Inc., 1998).

Canada Heirloom Series. *Visionaries: Canadian Triumphs, Vol. VI* (Mississauga, ON: Heirloom Publishing Inc., 1996).

Carpenter, Thomas. *Inventors: Profiles in Canadian Genius* (Camden East, ON: Camden Publishing House, 1990).

Downey, James, and Claxton, Lois. *Innovation: Essays by Leading Canadian Researchers* (Toronto: Key Porter Books, 2002).

Hacker, Carlotta. *Inventors* (Calgary: Weigl, 2000).

Hughes, Susan. *Canada Invents* (Toronto: Owl Books, 2002).

Humber, Charles, J., ed. *Canada: From Sea unto Sea* (Mississauga, ON.: The Loyalist Press, 1968).

Johnston, David, and Tom Jenkins. *Innovative Nation: How Canadian Innovators Made the World Ingenious, Smarter, Smaller, Kinder, Safer, Healthier, and Happier* (Toronto: Signal, 2017).

Mayer, Roy. *Inventing Canada: One Hundred Years of Innovation* (Vancouver: Raincoast Books, 1997).

Melady, John. *Breakthrough! Canada's Greatest Inventions and Innovations* (Toronto: Dundurn, 2013).

Nader, Ralph, Nadia Milleron, and Duff Conacher. *Canada Firsts* (Toronto: McClelland and Stewart, 1992).

McGoogan, Ken. *50 Canadians Who Changed the World* (Toronto: HarperCollins Publishers, 2013).

Nostbakken, Janis, and Jack Humphrey. *The Canadian Inventions Book: Innovations, Discoveries and Firsts* (Toronto: Greey de Pencer Publishers, 1976).

Schafer, D. Paul, *Celebrating Canadian Creativity* (Oakville, ON: Rock's Mills Press, 2016).

Spencer, Bev, and Bill Dickson. *Made in Canada: 101 Amazing Achievements* (Toronto: Scholastic Canada, 2003).

Trottier, Maxine. *Canadian Inventors* (Toronto: Scholastic Canada, 2004).

Van Ruskenveld, Yvonne and the Department of the Secretary of State. *About Canada: Innovation in Canada* (Ottawa: Ministry of Supply and Services, 1988).

Wojna, Lisa. *Canadian Inventors: Fantastic Feats and Quirky Contraptions* (Alberta: Folklore Publishers, 2004).

Wojna, Lisa. *Canadian Firsts: Inventions, Sports, Medicine, Space, Women's Rights, Explorers, Science, Research, Arts, World Affairs* (Edmonton: Folklore Publishers, 2008).

Index

The index includes only substantive references to people, inventions, discoveries, devices, and themes. Many other topics and persons are mentioned briefly in passing in the text.

Abbott, Maude, 2
Abortion, Canadian policy on, 266–267
Activism, Canadian contributions to, 152ff., 266–269
Actors, notable Canadian, 140
Adult education, see Education, adult
Advocacy, see Activism
Ahearn, Thomas, 86–87
Aircraft, 6–7, 34
Alouette 1, 37, 57
Analytic plotter, 40–41
Anik A1, 37
Animation, Canadian contributions to, 138
Anka, Paul, 4
Antigonish Movement, 166–168
Arbour, Louise, 252–253
Arcand, Denys, 141
Arendz, Mark, 230
Armstrong, Jim, 229
Artificial intelligence (AI), Canadian contributions to, 241–245
Arts administration, Canadian contributions to, 148ff., education in, 150–151
Arts, Canadian contributions to, 121ff.
Arts, investment in, 277
Atkins, George, 53
Atlas of Canada, The (Department of the Interior), 40
Atwood, Margaret, 5–6, 132
Automation, 85–86
Automatistes, 144–145
Avro Arrow, 6–7
Avro Jetliner, 6–7
Back, Frédéric, 7–8
Baldwin, Casey, 28
Ballet, 8–9
Banting, Frederick G., 9, 94–97

Barlow, Maude, 188–191
Basketball, Canadian contributions to, 9–10, 208ff.
Bata, Sonja, 10
Bata, Thomas John, Jr., 10
Beaverbrook, Lord (Max Aitken), 3–4
Belaney, Archibald, 177–179
Bell, Alexander Graham, 11, 27–28, 31, 45–46
Bell, Mabel, 31
Berton, Pierre, 11–12
Best, Charles H., 94–95
Bethune, Norman, 12, 104–106
Bigelow, Wilfred, 111–113
BlackBerry, 57
Bluenose, The, 12–13
Boat people, see Refugees
Bobsledding, Canadian contributions to, 228
Bombardier (company), 35–36
Bombardier, Joseph-Armand, 13, 34
Boucher, Gaëtan, 225
Bowling, 5-pin, 13
Breakey, Norman, 87
BRP, 36
Burton, Eli Franklin, 88
Burtynsky, Ed, 139
Cable, transatlantic and transpacific, 44–45
Callaghan, John, 112
Callaghan, Morley, 131
Canada Dry, see Ginger ale
Canadarm, 38
Canada-USSR 1972 hockey series, 204
Canadian Broadcasting Corporation, 52–53; International Service, 53–54
Canadian Opera Company, 126
Canadian Pacific Railway (CPR), 30–31
Cannabis, see Marijuana
Canoe, 19–20
Canola, 72–74
Carroll, Thomas, 75
Cartography, see Mapmaking
Champlain, Samuel de, 19
Charter of Rights and Freedoms, see Constitution, Canadian

Children's literature, Canadian contributions to, 134
Chiropractic, Canadian contributions to, 108ff.
Churchland, Patricia (Smith), 102
Churchland, Paul Montgomery, 102
Circus, Canadian contributions to, 147–148
Cirque du Soleil, 147–148
Coady, Moses, 166–168
Cobalt-60 bomb, 106–108
Collip, J.B., 94–95
Combine harvester, self-propelled, 75
Comedy, Canadian contributions to, 145ff.
Comic books, Canadian contributions to, 134–135
Confederacies, Indigenous, 245–246
Constitution, Canadian, 261–266; as model for other countries, 265
Cranston, Toller, 224–225
"Crazy Canucks," 227
Creativity, nature of Canadian, 15–16, 273–274
Creed, Frederick George, 55
Cuddy, Lola, 100
Cunard Line, ships of, 23–24
Cunard, Samuel, 21–24
Curling, Canadian contributions to, 217–222
Dagg, Anne Innis, 60–61
Dallaire, Roméo, 170–172
Darnell, Bill, 193
David Suzuki Foundation, 188
Desbarats, Georges-Edouard, 54–55
Diaphone, 26–27
Disabilities, Canadian contributions to assisting persons with, 153ff.
Disabled Peoples' International, 158–159
Distance education, see Education, distance
Documentary films, Canadian contributions to, 138–139
Doidge, Norman, 101–102
Drake, 123–125
Edmonton Grads, 10
Education, adult, 2–3, 239–240
Education, Canadian contributions to, 235ff.
Education, distance, 238–239
Edugyan, Esi, 133
Edwards, Henrietta Muir, 250
Electric light, invention of, 83–84
Electric oven, 87
Enns, Henry, 157–160
Entertainment, Canadian contributions to, 121ff.
Environmentalism, Canadian contributions to, 175ff.
Evans, Matthew, 83
Faith and Light, 156
Farm broadcasts, on radio, 52–53
Fenerty, Charles, 69–70
Fessenden, Reginald Aubrey, 50
Figure skating, see Skating
Film, Canadian contributions to, 135ff.
Fishing industry, 65ff.
Fitzpatrick, Alfred, 240
Fleming, Sandford, 42–43
Fog horn, steam, 26
Foster, Charles Basil, 137
Foster, David, 125
Foulis, Robert, 24–26
Fox, Terry, 160–163
Freer, James, 136
French fries, 75ff.
Frontier College, 240
Fur trade, 66ff.
Gesner, Abraham, 78–80
Ginger ale, 76–77
Gisborne, Frederick N., 44
Gosling, James, 57–58
Gould, Glenn, 128
Graphic novels, see Comic books
Greene, Nancy, 227
Greenpeace, 191–194
Greenwich Mean Time System, 43
Grey Owl, see Belaney, Archibald
Group of Seven, 142–144
Hadfield, Chris, 38–39
Haliburton, Thomas Chandler, 146
Hansen, Rick, 163–166
Harkin, James B., 232
Health care, 90ff.
Hébert, Anne, 131
Heggtveit, Anne, 226–227
Hein, David, 119
Helava, U.V., 40–41
Hinton, Geoffrey, 242–243
Hockey, Canadian contributions to, 200ff.

Hockey, international expansion of, 206–207
Hockey, women's participation in, 205–206
Holland, Andrew and George, 135–136
Holland, Samuel, 39–40
Hoodless, Adelaide, 240
Hopps, John, 112
Hospital for Sick Children, 110–111
Howard, Glenn, 218
Howlett, Les, 40
Human Rights Museum, 263–264
Humour, see Comedy
Humphrey, John, 251–252
Humphries, Kaillie, 228
Hunter, Bob, 191–192
Huntsman, Archibald, 66
Hydrofoil, 27–29
Ice wine, 77–78
IMAX, 140–141
Indigenous peoples, creative contributions of, 19, 65, 245
Innis, Harold, 58–61
Innis, Mary Quayle, 60
Innovation, Canadian, challenges to, 275–276; innovation clusters, 276–277
Instant replay, 208
Insulin, 9, 94–97
Iseler, Elmer, 127
Java programming language, 57–58
Jenkins, Clarence "Shorty," 219–221
Johannsen, Herman Smith "Jack Rabbit," 231
Johns, Harold, 106–108
Kayak, 19–20
Kerosene, use in lighting, 78–79
Kidd, J. Roby, 3, 240
Kielburger, Craig, 172–175
Kielburger, Marc, 173–174
Kingsbury, Mikaël, 227–228
Kirsch, Philippe, 253
Klassen, Cindy, 226
Klein, Naomi, 195–197
L'Arche, 154–157
Lachance, Fernand, 76
Lacrosse, Canadian contributions to, 215–216
Laurence, Margaret, 132
Le May Doan, Catriona, 225–226
Leaver, Eric, 84–85

Leggo, William, 54–55
Lewis, Stephen, 168–170
Literature, Canadian contributions to, 130ff.
Logan, William, 40
Lumber industry, 68ff.
Macleod, J.J.R., 94–95, 97
Maillet, Antonine, 132
Mak, Tak Wah, 115–116
Man in Motion tour, 164
Mapmaking, 39ff.
Marathon of Hope, 161–163
Marconi, Guglielmo, 49–50
Marijuana, Canadian policy on, 269
Maritimes, and shipbuilding, 20ff.
Marquis wheat, 70–72
Martel, Yann, 132–133
Mau, Sandra, 244
Mayer, Louis B., 137
McCain Frozen Foods, founding and growth of, 75–76
McClung, Nellie, 250
McCulloch, Ernest, 113–114
McCulloch, Jack, 225
McCurdy, J.A. Douglas, 31
McKeever, Brian, 229, 230
McKinney, Louise, 250
McLaren, Norman, 7–8, 138, 139
McLaughlin, John J., 76–77
McLeod, Alistair, 133
McLuhan, Marshall, 61–63
McMaster, Elizabeth, 110–111
McTaggart, David, 192–193
Medically assisted dying, Canadian policy on, 268
Medicine, 90ff.
Milner, Brenda, 100
Miner, Jack, 175–177
Mitchell, W.O., 131
Moir, Scott, 223–224
Montgomery, Lucy Maud, 4–5
Morgentaler, Henry, 267
Morse keyboard perforator, 55
Mounce, G.R., 85
Mowat, Farley, 180–182
Multiculturalism, 237ff., 255–261
Munro, Alice, 133–134
Murphy, Emily, 250
Music, Canadian contributions to, 121ff.
Mustard, Fraser, 240–241, 242
Naismith, James, 9–10, 208–209

National Film Board, 138–139
Neuroscience, Canadian contributions to, 98ff.
Newspapers, Canadian contributions to, 54ff.
Noorduyn, Bob, 33
Northey, John Pell, 26–27
Oil Springs, Ontario, 80–81
Olcott, Sidney, 136
Ondaatje, Michael, 132
Osler, William, 91–94, 101
Ouimet, Ernest, 136
Oxyacetylene torch, 83
Pablum, 111
Pacemaker, heart, 112–113
Paint roller, 87–88
Palmer, Daniel David, 108–110
Paralympic Games, Canadian contributions to, 229–230
Parks Canada, 232–233
Parlby, Irene, 250
Patch, John, 24–25
Peacekeeping, Canadian contributions to, 253–255
Pearson, Lester, 254
Pellan, Alfred, 144
Penfield, Wilder, 98–100
Peterson, Oscar, 129
Petrochemical industry, 78ff.
Photographs, halftone, 54–55
Politics, Canadian contributions to, 245ff.
Politics, Canadian, female pioneers in, 250–251
Poutine, 76
Propeller, screw, 25
Propeller, variable pitch, 32–33
Pulp and paper industry, 69ff.
Quebec Act, 246–248
Radio, Canadian contributions to development of, 49ff.
Raptors, Toronto, 213–214
Recreation, Canadian contributions to, 230–233
Refugees, Canadian response to, 117ff.; Vietnamese, 117–118; Syrian, 118
Research in Motion, founders of, 57
Richardson, Ernie, 218
Richler, Mordecai, 132
Robinson, W.A., 30
Rogers, Edward Samuel (Ted), 52
Roy, Gabrielle, 131
Rubinstein, Louis, 223
Ruttan, Henry, 30
Ryan, Thomas F., 13
Sabine, Edward, 40
Saint-Jacques, David, 38
Same-sex marriage, Canadian policy on, 267–268
Sankoff, Irene, 119
Satellite communications, 57
Saunders, Charles E., 70–72
Schafer, R. Murray, 129–130
Schmirler, Sandra, 218
Scott, Barbara Anne, 223
Sellers, Margaret (Peg), 228
Selye, Hans, 102–104
September 11 terrorist attacks, response to, in Gander, 119–120
Seton, Ernest Thompson, 179–180
Sharp, Samuel, 30
Shields, Carol, 133
Shore, Howard, 140
Shuster, Joe, 134
Sicard, Arthur, 36
Silver Dart, 31
Skating, Canadian contributions to, 222–226
Ski-Doos, see Snowmobiles
Skiing, Canadian contributions to, 226–228
Sleeper car, 30
Snowblower, 36–37
Snowmobiles, 13, 34–36
Soundscapes, 129–130
Space flight, Canadian contributions to, 37ff.
Spar Aerospace, 37
Speed skating, see Skating
Sports, Canadian contributions to, 198ff.
Standard time, 42–43
Staples thesis, 58
Statute of Westminster, Canadian role in formulating, 249–250
Stem cell research, 113–116
Stephenson, William, 55–56
Strong, Maurice, 182–186
Sutton, Richard, 243
Suzuki, David, 186–188
Swimming, synchronized, Canadian contributions to, 228
Tapscott, Don, 63–64

Telephone, invention and development of, 45ff.
Thompson, David, 20
Thomson, John, 70
Thomson, Tom, 142–144
Till, James, 113–114
Time-keeping, 41ff.
Toronto International Film Festival (TIFF), 142
Trans Canada Trail, 233
Transmission electron microscope, 88–89
Transportation, by air, 31ff.
Transportation, by rail, 29ff.
Transportation, by water, 18ff.
Transportation, by wheeled vehicles, 37
Tripp, Charles N., 80
Trudeau, Pierre Elliott, 261–262
Turnbull, Wallace Rupert, 32–33
Vanier, Jean, 153–157
Virtue, Tessa, 223–224
Visual arts, Canadian contributions to, 142ff.
Waldo, Carolyn, 228
War of 1812, 248
Warner, Jack, 137
Watson, Paul, 193–194
We organizations, 173–175
Werenich, Ed "the Wrench," 219
Wheeler, Lucile, 226
Williams, James Miller, 80–81
Willson, Thomas "Carbide," 82–83, 84
Women's Institute, 240
Woodward, Henry, 83
Woolstencroft, Lauren, 229
World War I, Canada's role in, 248–249

www.ingramcontent.com/pod-product-compliance
Lightning Source LLC
Chambersburg PA
CBHW071806080526
44589CB00012B/702